Vishal S. Bagul, Rakesh E Mutha, Sanjaykumar B Bari
Mucilage and Gums

Also of interest

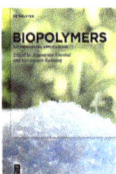

Biopolymers.
Environmental Applications
Edited by: Jeyaseelan Aravind and Murugesan Kamaraj, 2023
ISBN 978-3-11-099872-6; e-ISBN (PDF) 978-3-11-098718-8

Biopolymer Conjugates.
Industrial Applications
Edited by: Swati Sharma and Ashok Kumar Nadda, 2024
ISBN 978-3-11-078576-0; e-ISBN 978-3-11-078583-8

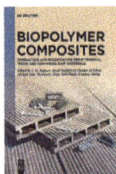

Biopolymer Composites.
Production and Modification from Tropical Wood and Non-Wood Raw
Materials
Edited by: Salit Mohd Sapuan, Syeed SaifulAzry Osman Al Edrus, Ahmad
Adlie Shamsuri, Aizat Abd Ghani and Khalina Abdan, 2023
ISBN: 978-3-11-162237-8; e-ISBN 978-3-11-076922-7

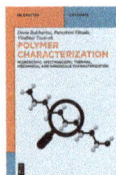

Polymer Characterization.
Microscopic, Spectroscopic, Thermal, Mechanical and Nanoscale
Characterization
Daria Bukharina, Paraskevi Flouda and Vladimir Tsukruk, 2025
ISBN: 978-3-11-134536-9; e-ISBN 978-3-11-134574-1

Polymers.
Chemistry, Morphology, Characterization, Processing, Technology and
Recycling
Mohamed Elzagheid, 2025
ISBN: 978-3-11-158565-9; e-ISBN 978-3-11-158573-4

Vishal S. Bagul, Rakesh E Mutha,
Sanjaykumar B Bari

Mucilage and Gums

Natural Polymers, Biodegradability, and Sustainable
Development

DE GRUYTER

Authors
Vishal Suresh Bagul
H. R. Patel Institute of Pharmaceutical Education and Research
Shirpur, Dhule 425 405
Maharashtra, India
Bagulvishal1@gmail.com

Dr. Rakesh Eshwarlal Mutha
H. R. Patel Institute of Pharmaceutical Education and Research
Shirpur, Dhule 425 405
Maharashtra, India
rakeshmutha123@yahoo.co.in

Dr. Sanjaykumar Baburao Bari
H. R. Patel Institute of Pharmaceutical Education and Research
Shirpur, Dhule 425 405
Maharashtra, India
drsbbari@gmail.com

ISBN 978-3-11-167316-5
e-ISBN (PDF) 978-3-11-167350-9
e-ISBN (EPUB) 978-3-11-167450-6

Library of Congress Control Number: 2025951340

Bibliographic information published by the Deutsche Nationalbibliothek
The Deutsche Nationalbibliothek lists this publication in the Deutsche Nationalbibliografie;
detailed bibliographic data are available on the Internet at http://dnb.dnb.de.

© 2026 Walter de Gruyter GmbH, Berlin/Boston, Genthiner Straße 13, 10785 Berlin
Cover image: ligora/iStock/Getty Images Plus
Typesetting: Integra Software Services Pvt. Ltd.

www.degruyterbrill.com
Questions about General Product Safety Regulation:
productsafety@degruyterbrill.com

Preface

The growing global emphasis on sustainability and eco-conscious innovation has led to renewed interest in natural materials that harmonize with both technological advancement and environmental responsibility. Among these, mucilage and gums have emerged as remarkable natural polymers offering a spectrum of applications across pharmaceutical, biomedical, food, and industrial sectors. Their unique structural, functional, and rheological properties, combined with inherent biodegradability and biocompatibility, position them as vital contributors to the advancement of green chemistry and sustainable materials science.

This book brings together current knowledge, innovations, and research trends focused on the chemistry, extraction, characterization, modification, and multifaceted applications of mucilage and gums. The chapters have been carefully curated to reflect both fundamental principles and applied research, emphasizing the transition from laboratory findings to real-world industrial applications.

The book is intended to serve as a valuable resource for researchers, academicians, industry professionals, and students engaged in the fields of natural product chemistry, polymer science, pharmaceutics, food technology, and biomaterials. It underscores the role of these natural biopolymers not merely as substitutes for synthetic polymers but as key elements in fostering sustainable development and circular economy principles.

We sincerely express our gratitude to Jessika Kischke, De Gruyter Publication House, for recognizing the scientific relevance of this work and for their continued commitment to disseminating high-quality scholarly publications. Finally, we dedicate this book to all those who continue to pursue scientific inquiry with integrity, curiosity, and a vision for a sustainable future. Additionally, we sincerely request readers to send their valuable suggestions and constructive comments for improvement in the next edition of the book.

Vishal Bagul
Rakesh Mutha
Sanjaykumar Bari

https://doi.org/10.1515/9783111673509-202

Contents

The Author's Bio

Mr. Vishal Suresh Bagul Born in 1992, he is a Ph.D. scholar at H. R. Patel Institute of Pharmaceutical Education and Research, Shirpur, India, focusing on the exploration of natural polymers and their application in pharmaceutical sciences.

Dr. Rakesh Eshwarlal Mutha born in 1982, is Associate Professor of Pharmacognosy with a focus on isolation, characterization and modification of natural polymers at H. R. Patel Institute of Pharmaceutical Education and Research, Shirpur, India.

Dr. Sanjaykumar B Bari born in 1971, is Principal and Professor of Pharmaceutical Chemistry with a focus on Synthesis and modification of natural polymers at H. R. Patel Institute of Pharmaceutical Education and Research, Shirpur, India.

https://doi.org/10.1515/9783111673509-204

1 Overview of Natural Polymers

1.1 Introduction to Natural Polymers

Natural polymers are biocompatible, biodegradable, and renewable macromolecules derived from plants, animals, or microbial sources [1]. These polymers have been widely explored for their applications in pharmaceuticals, biomedical engineering, food industries, and environmental sustainability. Their unique physicochemical properties, including biodegradability, non-toxicity, and high functionality, make them valuable alternatives to synthetic polymers. Natural polymers are high-molecular-weight compounds that are characterized by an impressive array of remarkable properties, which include not only toughness and unparalleled durability but also exceptional transparency and impressive flexibility [2]. These indispensable features render them as unique and versatile materials that can be utilized across a wide spectrum of applications in numerous diverse industries, effectively transforming how we approach materials science. Most of the commercially available polymers that we encounter in our daily lives today are derived from finite and limited natural resources. The industrial processes employed for their production, however, often result in the release of toxic pollutants into the environment, raising significant concerns about sustainability and ecological impact. In stark contrast, natural polymers stand out, as they are not only highly abundant but also renewable. Some sources, such as natural rubber, have been cultivated and harvested for commercial use for many thousands of years, which highlight their long-standing value in various human endeavors and industries and underscores the need to shift our focus toward more sustainable practices that harness these natural materials.

This ongoing search for more eco-friendly options has involved significant efforts to not only improve upon existing natural polymers but also to produce entirely new products that are sourced from various natural materials [3]. This has led to the generation and development of innovative types of natural polymers that are not only effective but also align with the principles of sustainability and ecological responsibility [4]. The principal reasons for the increasing focus on studying and utilizing natural polymers are manifold. Firstly, these materials are inherently biocompatible, making them ideal candidates for a broad spectrum of medical, pharmaceutical, and biotechnological applications. For instance, they are commonly used as scaffolds or matrices for implants, intricate skin care systems, carriers designed for drugs and various biological molecules, as well as stabilizers in both food and pharmaceutical products.

Secondly, many natural polymers are characterized by the presence of functional groups that can form reversible chemical bonds with inorganic fillers, semiconductors, or metallic particles, as well as carbon-based nanomaterials and enzymes [5]. This characteristic leads to the development of composites with significantly enhanced electrical and mechanical properties.

https://doi.org/10.1515/9783111673509-001

Thirdly, natural polymers demonstrate the ability to bind water, are readily available from various sources, and are generally more cost-effective than their synthetic counterparts. Additionally, they are frequently biodegradable and, in some specific cases, can even be considered edible. This combination of attributes makes them particularly appealing in a world that increasingly values sustainability and health-conscious options.

These polymers are primarily organic compounds and are typically formed through processes like polymerization, where smaller molecules join together to form a larger, complex structure. Unlike synthetic polymers, which are artificially produced from petrochemical resources, natural polymers are derived from renewable resources and are environmentally friendly [6].

Natural polymers can be classified into two broad categories: biodegradable and nonbiodegradable. Biodegradable polymers, such as starch and cellulose, can be broken down by microorganisms and enzymes in the environment. Nonbiodegradable polymers, like rubber, may take longer to decompose but are still naturally derived (Figure 1.1). These polymers are integral to various biological functions, including providing structure to cells, storing energy, and serving as catalysts for biochemical reactions [7].

1.2 Types of Natural Polymers

1.2.1 Polysaccharides

Polysaccharides are carbohydrates composed of long chains of monosaccharide units. These molecules perform diverse biological functions, including energy storage, structural integrity, and signaling. The primary polysaccharides found in nature include starch, cellulose, glycogen, and chitin:

– **Starch**: Starch is a plant-derived polymer composed of glucose units. It is a primary energy storage material in plants and is found in foods like potatoes, rice, and corn. Starch consists of two molecules: amylose (a linear polymer) and amylopectin (a branched polymer). Starch is widely used in food processing, as well as in industrial applications for producing biodegradable plastics.
– **Cellulose**: Cellulose is the most abundant natural polymer on Earth and forms the structural component of plant cell walls. It is made up of repeating glucose units linked by β-1,4-glycosidic bonds. Cellulose provides mechanical strength to plants and is crucial for maintaining their shape and rigidity. It is also used in the textile industry to make fabrics like cotton and linen. Cellulose can be processed into paper, biofuels, and biodegradable plastics.
– **Glycogen**: Glycogen is a branched polysaccharide that serves as the primary storage form of glucose in animals, particularly in the liver and muscles. When the body needs energy, glycogen is broken down into glucose molecules, which are

Natural Polymers		
Natural Polymers of Plant Origin	Polysaccharides (Cyclodextrins, Cellulose, Hemicellulose, Starch, Inulin, Pectin, Glucomannan, Guar Gum, Arabinogalactan)	
	Proteins (Soy Protein)	
	Polyesters (From Higher Plants)	
Natural Polymers of Animal Origin	Polysaccharides (Chitosan, Chondroitin, Chondroitin, Sulphate)	
	Proteins (Collagen, Gelatin, Albumin, Fibrin, Silk Fibroin)	
	Resin (Shellac)	
Natural Polymers of Microbial Origin	Polysaccharides (Alginate, Dextran)	
	Polyesters (Phas)	
	Polyamides (Poly-glutamate)	
	Polyanhydrides (Polyphosphate)	

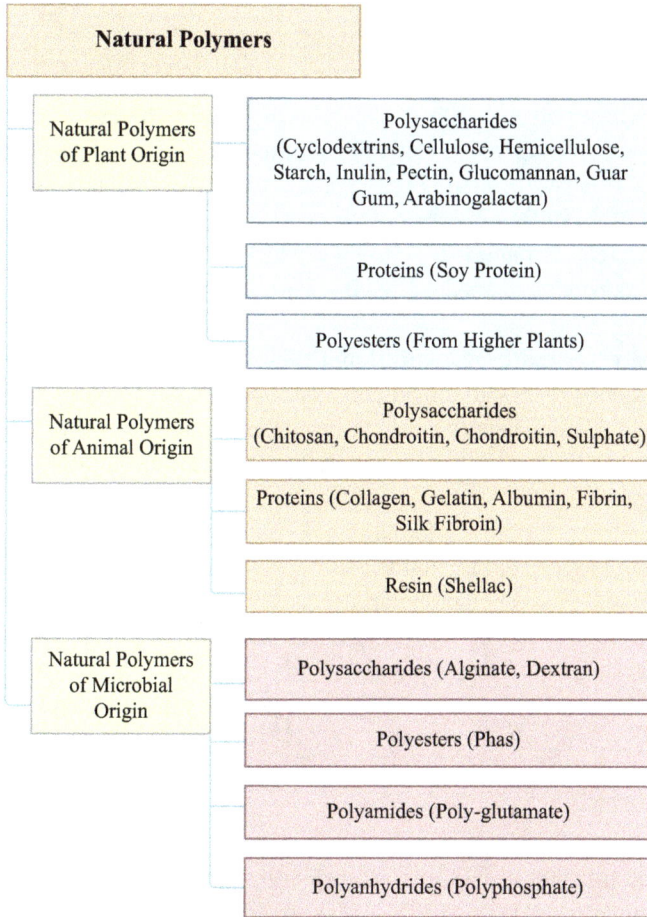

Figure 1.1: Natural polymers [7, 8].

then used to produce ATP. It has a similar structure to amylopectin but is more extensively branched.

– **Chitin**: Chitin is a structural polysaccharide found in the exoskeletons of arthropods (such as insects and crustaceans) and in the cell walls of fungi. It is made up of *N*-acetylglucosamine units and is similar to cellulose in structure. Chitin can be processed into chitosan, a derivative with several industrial and biomedical applications, including wound dressings and as a biodegradable plastic [7, 8].

1.2.2 Proteins

Proteins are another class of natural polymers, made up of amino acids linked together by peptide bonds. These polymers play crucial roles in nearly all biological processes, from providing structural support to catalyzing biochemical reactions:

– **Collagen**: Collagen is a fibrous protein found in connective tissues, such as skin, bones, and cartilage. It provides tensile strength and elasticity to tissues and is the most abundant protein in mammals. Collagen is widely used in the cosmetics industry (for antiaging products), medical fields (in wound healing), and in the food industry.
– **Silk**: Silk is a protein produced by silkworms and spiders. The primary component of silk is fibroin, a protein with a highly ordered, crystalline structure. Silk fibers are known for their tensile strength, smooth texture, and sheen. Silk has been used for thousands of years to make clothing and textiles, and it also has biomedical applications due to its biocompatibility.
– **Keratin**: Keratin is a structural protein found in hair, nails, and skin. It provides rigidity and protection to these tissues and is a key component of the epidermis. Keratin is also used in the production of materials like leather and wool.
– **Elastin**: Elastin is a protein found in connective tissues that provides elasticity. It allows tissues such as skin, lungs, and blood vessels to stretch and return to their original shape. Elastin is crucial for maintaining tissue flexibility and resilience.

1.2.3 Nucleic Acids

Nucleic acids are natural polymers composed of nucleotide units. These polymers store and transmit genetic information and are crucial for cellular functions such as protein synthesis:

– **DNA (Deoxyribonucleic Acid)**: DNA is a double-stranded polymer that carries genetic information in all living organisms. It is composed of nucleotide monomers, each consisting of a sugar molecule (deoxyribose), a phosphate group, and a nitrogenous base (adenine, thymine, cytosine, or guanine). The sequence of these nucleotides encodes the genetic blueprint for the development, function, and reproduction of organisms.
– **RNA (Ribonucleic Acid)**: RNA is similar to DNA but is typically single-stranded. It plays a crucial role in the synthesis of proteins by acting as a messenger between the DNA in the nucleus and the ribosomes in the cytoplasm. RNA is composed of ribose sugar, phosphate groups, and nitrogenous bases (adenine, uracil, cytosine, and guanine).

1.2.4 Natural Rubber

Natural rubber, or latex, is a polymer made from the monomer isoprene, which is obtained from the sap of rubber trees (*Hevea brasiliensis*). The polymer is formed by the polymerization of isoprene molecules, resulting in a long-chain elastomer. Rubber is highly elastic and is used in the production of a wide range of products, including tires, footwear, medical gloves, and various industrial goods [9].

1.2.5 Lignin

Lignin is a complex polymer found in the cell walls of plants, particularly in woody plants. It provides rigidity and resistance to decay and plays a crucial role in strengthening the plant structure. Lignin is primarily composed of phenylpropanoid units and is responsible for the structural integrity of vascular tissues in plants. It is a critical component of wood, and its biodegradability makes it a focus of research in producing sustainable materials.

1.3 Classification of Natural Polymers [8]

Natural polymers can be classified based on their origin, composition, and functionality. The primary classification categories include:

1.3.1 Polymers from Plants (Plant-Based Polymers)

These are natural polymers sourced from plants, and they form the bulk of natural polymers used in various applications. These polymers are mainly polysaccharides (carbohydrates) and proteins. Examples include starch, cellulose, and rubber.

1.3.2 Polymers from Animals (Animal-Based Polymers)

Animal-derived polymers are typically proteins and nucleic acids. These polymers serve essential functions in the structure and regulation of animal tissues. Examples include collagen, silk, and chitin.

1.3.3 Polymers from Microorganisms (Microbial Polymers)

Microorganisms such as bacteria and fungi can produce natural polymers. These polymers can be polysaccharides or proteins and are often used in industrial applications
due to their unique properties. Examples include bacterial cellulose and polyhydroxyalkanoates (PHAs).

1.3.4 Biopolymers (Based on Biochemical Composition)

Biopolymers are natural polymers that include both synthetic and naturally occurring
types. These biopolymers can be further categorized into three broad groups: polysaccharides, proteins, and nucleic acids.

1.4 Extraction and Processing of Natural Polymers

Four fundamental processing methods are utilized in the production of natural polymers, depending on the origins of the materials employed. To elaborate further, in
the first instance, the method of directly extracting biopolymers from either vegetable
or animal sources stands out. This approach necessitates the purification of the extracted biopolymers, particularly when they are intended for use in applications that
demand a high degree of purity. The extraction process of biopolymers typically involves the treatment of the chosen raw materials with dilute acid and alkaline solutions. This treatment serves to reduce the presence of pigments, lipids, and both hydrophobic and hydrophilic substances, although it can inadvertently lead to damage
in the polymer chains Following this initial step, a concentrated alkaline solution is
applied to facilitate the removal of hemicellulose and lignin, which are often undesirable components in the final product. During the acid hydrolysis process, low-molar
mass products are generated, while alkaline treatment can lead to the unwanted de-
polymerization of cellulose, presenting yet another challenge. Nonetheless, the extraction of certain natural polymers can pose significant difficulties, as the application of
both acids and bases requires meticulous care [9].

This precaution is vital to maintain the structural integrity of sensitive compounds such as cellulose and chitosan, safeguarding them from any detrimental alter-
ations that could compromise their properties. Therefore, to render natural polymers
commercially viable, it is essential to use cost-effective starting materials in conjunction with efficient extraction processes that fulfill these stringent requirements.

Furthermore, other treatments may be employed, including the bleaching action
of oxidizing agents and hypochlorite, which play a crucial role in removing residual
color and lignin. Notably, the production of high molar mass natural polymers, which

tend to exhibit lower water solubility or heightened water absorption capabilities, can be achieved by adjusting various parameters.

These may include modifications to the type of solvent used, its concentration, the physical state in which the solvent or cosolvent is applied, and consideration of molecular size, all of which significantly influence the properties of the final product [9, 10].

1.4.1 Isolation Techniques

Isolation typically occurs after the initial removal of gross nonfibrous materials from the source, which can include parts such as leaf, stem, or trunk bark. Subsequently, the body of the specified part is typically ground to effectively liberate the fiber bundles. The process of fiber removal is often significantly facilitated by a range of various physical treatments, which may include methods such as maceration, pulping, or thorough washing. Additionally, several chemical actions that can be employed involve boiling or soaking the raw material extensively in different solvents, as well as steam treatment or the application of alkali in order to destroy the binding material, or alternatively, to cause the raw material to swell [9].

There are optional mechanical treatments that can be implemented, such as manual cleaning and brushing to carefully remove leaf epidermis or lumen material. However, it is important to note that this manual process can be minimized effectively by controlled growth and strategic harvesting of the plant before the production of leaves starts.

In addition, procedures like washing and the manipulation of adhesive conditions of fibers found within the lumen are utilized to efficiently extract fiber bundles. During the initial stage of this elaborate process, the cells that occur in strands or fascicles usually break along the sides of those strands, thereby releasing individual cells. In the subsequent stage, these individual cells then separate longitudinally. At times, particularly when an alkali treatment is utilized to remove the cellular adhesive material with the aim of enhancing mechanical properties, microfibrils along with microfibril radicals can remain stably bound together; essentially, microfibrils cling together much like the fibers themselves. On reaching this specific stage, all the cells of a type in combination are methodically separated into distinct individual cells [9].

Following that, refined fiber that has been treated with alkali and subjected to beating can be utilized to effectively improve the bonding of the microfibrils along with the radicals involved. The final refining stage, which may sometimes involve exposing the fibers to an ultrasound treatment, also results in the dissociation of single cells ultimately enhancing the structural integrity. If the fibrils are meticulously treated on their surface to form a fibrillated structure, this crucial process is completed in what is referred to as a fibrillization stage.

1.5 Role of Natural Polymers in Sustainable Development

In the age of rapid industrialization and growing environmental concerns, sustainability has become a key priority across multiple sectors. The demand for materials that are not only functional but also eco-friendly has led to an increasing interest in natural polymers. These polymers, derived from renewable sources like plants, animals, and microorganisms, play a crucial role in promoting sustainability by offering alternatives to synthetic polymers, which often have detrimental effects on the environment [10].

1.5.1 Biodegradability and Waste Management

One of the most significant contributions of natural polymers to sustainable development is their biodegradability. Unlike synthetic polymers such as polyethylene or polypropylene, which can persist in the environment for hundreds of years, natural polymers decompose naturally in the presence of microorganisms, water, and oxygen [11]. This characteristic makes them ideal for use in reducing environmental pollution, particularly plastic waste.

For example, polylactic acid (PLA), a biopolymer derived from renewable sources such as corn starch or sugarcane, has become a prominent substitute for conventional plastic. PLA is not only biodegradable but also compostable, which means it can break down into harmless organic compounds when disposed of correctly. This stands in stark contrast to petroleum-based plastics, which clog landfills, pollute oceans, and pose a significant threat to wildlife. Additionally, natural polymers like chitosan (derived from the shells of crustaceans) have found applications in the food industry for biodegradable packaging materials, offering a more sustainable alternative to plastic wraps and containers [12].

In waste management systems, natural polymers are increasingly being integrated into composting processes to reduce landfill waste. By incorporating biopolymers like starch-based plastics into packaging, single-use products, and agricultural films, the waste stream can be significantly reduced, making a substantial contribution to the circular economy [13].

1.5.2 Renewable Resource Base

Another crucial aspect of natural polymers is that they come from renewable resources. Unlike petroleum-based polymers, which rely on fossil fuels that are finite and polluting, natural polymers can be sourced from plants, animals, and microorganisms that are replenishable. This makes natural polymers an integral part of sustainable development strategies that focus on resource conservation and the reduction of carbon footprints [14].

For instance, cellulose, the most abundant organic polymer on Earth is sourced from wood, cotton, and other plant materials. With proper land management practices, cellulose can be continuously harvested from forests and agricultural crops. It has applications in textiles, paper, and packaging materials. The use of cellulose in place of synthetic fibers, such as polyester or nylon, can significantly reduce the environmental impact associated with textile production, including water consumption, energy usage, and chemical waste. Additionally, the cultivation of plants like hemp and flax, which provide natural fibers, offers an environmentally friendly alternative to cotton and synthetic fibers. These plants require fewer pesticides and fertilizers, making them less resource intensive [15].

Another example is natural rubber, which is sourced from the latex of rubber trees. Unlike synthetic rubber, which is derived from petroleum, natural rubber is renewable and biodegradable. As demand for eco-friendly products increases, the use of natural rubber in tires, footwear, and other industrial applications can significantly contribute to reducing the carbon footprint associated with rubber production [16].

1.5.3 Reducing Toxicity and Environmental Impact

Many synthetic polymers contain harmful chemicals, such as phthalates, bisphenol A (BPA), and styrene, which pose risks to human health and the environment. These chemicals can leach into the environment, contaminating water supplies, soil, and wildlife. In contrast, natural polymers are often nontoxic and biodegradable, reducing the risk of chemical pollution. For example, **chitin** and **chitosan**, derived from the exoskeletons of crustaceans, are used in biomedical and agricultural applications [12, 17]. These natural polymers are biocompatible, nontoxic, and have antimicrobial properties, making them safer alternatives to synthetic chemicals used in wound healing, drug delivery, and agricultural pest control.

Additionally, natural polymers like **gelatin**, derived from animal collagen, are used in the pharmaceutical and food industries. Gelatin is biodegradable and has minimal environmental impact compared to synthetic alternatives, which can accumulate in the environment and affect ecosystems. The adoption of natural, nontoxic polymers in various industries reduces the need for harmful chemical additives, thereby decreasing overall environmental toxicity [18].

1.5.4 Agricultural Sustainability

Natural polymers also play a key role in promoting sustainable agriculture. Biodegradable films made from natural polymers, such as **starch-based plastics** and **PLA,** are being used as mulch films in agricultural practices. These films provide an eco-friendly alternative to traditional plastic mulch, which often ends up in landfills or waterways [19]. The use of biodegradable films reduces plastic waste and eliminates the need for labor-intensive film removal after the growing season.

Additionally, **chitosan**, derived from chitin, has shown promise in agricultural applications as a natural pesticide and soil conditioner [20]. Chitosan can help reduce the need for harmful synthetic pesticides and fertilizers, thereby minimizing chemical runoff into water bodies and reducing soil contamination. By promoting the use of biopolymers in agriculture, we can improve soil health, reduce environmental contamination, and support the long-term sustainability of agricultural practices [19, 21].

1.5.5 Biofuels and Energy Production

The production of biofuels from natural polymers is another area where these materials contribute to sustainable development. **Cellulose** and other plant-based biopolymers can be converted into biofuels, such as ethanol and butanol, through processes like fermentation and hydrolysis. Unlike fossil fuels, which release large amounts of carbon dioxide and other greenhouse gases into the atmosphere, biofuels derived from natural polymers can offer a more sustainable energy solution by reducing reliance on petroleum and mitigating climate change [22].

For instance, **bioethanol** is produced from the fermentation of sugars present in natural polymers like starch and cellulose. This biofuel is used as an alternative to gasoline, reducing the carbon footprint of transportation. The growing interest in **second-generation biofuels** (derived from lignocellulosic biomass) further enhances the role of natural polymers in sustainable energy production. By utilizing waste biomass, such as agricultural residues, for biofuel production, natural polymers can contribute to waste reduction and energy sustainability [23].

1.5.6 Advancements in Green Technologies

In addition to their direct applications, natural polymers are increasingly being incorporated into innovative green technologies. For example, **nanocellulose**, a form of cellulose that has been processed to produce nanoscale fibers, has shown promise in applications such as water filtration, energy storage, and lightweight composites. Nanocellulose-based materials are strong, lightweight, and biodegradable, making them ideal for use in environmentally friendly packaging and construction materials [24].

Moreover, **bio-based adhesives** derived from natural polymers such as starch, protein, or cellulose are replacing synthetic, petroleum-based adhesives in various industries. These adhesives are biodegradable and nontoxic, offering a sustainable alternative in the construction, automotive, and packaging sectors [25].

1.6 Economic and Social Impact

The use of natural polymers in sustainable development also extends beyond environmental benefits to economic and social dimensions. The cultivation and processing of natural polymers can create new industries and job opportunities, especially in rural areas where agricultural products like rubber, flax, and hemp are grown. By promoting the use of locally sourced materials, sustainable industries based on natural polymers can contribute to economic development while reducing reliance on imported synthetic materials [26].

Moreover, the growing demand for sustainable products derived from natural polymers can lead to the creation of new markets and innovations. This, in turn, stimulates research and development, leading to the production of more efficient and eco-friendly materials. Socially, the shift toward natural polymers can promote greater awareness of environmental sustainability and inspire consumers to make more informed choices [27].

References

[1]   Varghese, S. A., Rangappa, S. M., Siengchin, S., & Parameswaranpillai, J. (2020). Natural polymers and the hydrogels prepared from them. In Hydrogels based on natural polymers (pp. 17–47). Elsevier.
[2]   John, M. J., & Thomas, S. (Eds.). (2012). Natural polymers: volume 1: composites (Vol. 16). Royal society of chemistry.
[3]   Amara, A. A. A. F. (2022). Natural polymer types and applications. Biomolecules from natural sources: advances and applications, 31–81.
[4]   Vikhareva, I. N., Buylova, E. A., Yarmuhametova, G. U., Aminova, G. K., & Mazitova, A. K. (2021). An overview of the main trends in the creation of biodegradable polymer materials. Journal of chemistry, 2021(1), 5099705.
[5]   Olatunji, O. (2016). Natural polymers (pp. 1–17). Cham, Switzerland: Springer International Publishing.
[6]   Causin, V. (2015). Polymers on the crime scene. In Polymers on the Crime Scene: Forensic Analysis of Polymeric Trace Evidence (pp. 105–166). Cham: Springer International Publishing.
[7]   Bhatia, S. (2016). Natural polymers vs synthetic polymer. In Natural polymer drug delivery systems: nanoparticles, plants, and algae (pp. 95–118). Cham: Springer International Publishing.
[8]   Pillai, C. K. S. (2010). Challenges for natural monomers and polymers: novel design strategies and engineering to develop advanced polymers. Designed Monomers and Polymers, 13(2), 87–121.
[9]   Elias, H. G. (Ed.). (2012). Macromolecules 1: Volume 1: Structure and Properties. Springer Science & Business Media.

[10] Balaji, A. B., Pakalapati, H., Khalid, M., Walvekar, R., & Siddiqui, H. (2017). Natural and synthetic biocompatible and biodegradable polymers (Vol. 286, pp. 3–32). Amsterdam, The Netherlands: Elsevier.

[11] Silva, A. C., Silvestre, A. J., Vilela, C., & Freire, C. S. (2021). Natural polymers-based materials: A contribution to a greener future. Molecules, 27(1), 94.

[12] Olatunji, O. (2016). Natural polymers (pp. 1–17). Cham, Switzerland: Springer International Publishing.

[13] Rayner, M., Östbring, K., & Purhagen, J. (2015). Application of natural polymers in food. Natural polymers: Industry techniques and applications, 115–161.

[14] Bassas-Galia, M., Follonier, S., Pusnik, M., & Zinn, M. (2017). Natural polymers: a source of inspiration. In Bioresorbable polymers for biomedical applications (pp. 31–64). Woodhead Publishing.

[15] Saldívar-Guerra, E., & Vivaldo-Lima, E. (2013). Introduction to polymers and polymer types. Handbook of polymer synthesis, characterization, and processing, 1–14.

[16] Shubhra, Q. T., Alam, A. K. M. M., Gafur, M. A., Shamsuddin, S. M., Khan, M. A., Saha, M., . . . & Ashaduzzaman, M. (2010). Characterization of plant and animal based natural fibers reinforced polypropylene composites and their comparative study. Fibers and Polymers, 11(5), 725–731.

[17] Sapuan, S. M., Azhari, C. H., & Nurazzi, N. M. (Eds.). (2024). Polymer Composites Derived from Animal Sources. Elsevier.

[18] Babu, R. J., Annaji, M., Alsaqr, A., & Arnold, R. D. (2019). Animal-based materials in the formulation of nanocarriers for anticancer therapeutics. In polymeric nanoparticles as a promising tool for anti-cancer therapeutics (pp. 319–341). Academic Press.

[19] Rodríguez-Hernández, Juan. "Polymers against microorganisms." Polymers against Microorganisms: On the Race to Efficient Antimicrobial Materials. Cham: Springer International Publishing, 2016. 1–11.

[20] Dawes, E. A. (Ed.). (2012). Novel biodegradable microbial polymers (Vol. 186). Springer Science & Business Media.

[21] Kawai, F. (2006). Breakdown of plastics and polymers by microorganisms. Microbial and eznymatic bioproducts, 151–194.

[22] Walton, A. (2012). Biopolymers. Elsevier.\

[23] George, A., Sanjay, M. R., Srisuk, R., Parameswaranpillai, J., & Siengchin, S. (2020). A comprehensive review on chemical properties and applications of biopolymers and their composites. International journal of biological macromolecules, 154, 329–338.

[24] Mazuki, N. F., Saadiah, M. A., Fuzlin, A. F., Khan, N. M., & Samsudin, A. S. (2022). Basic aspects and properties of biopolymers.

[25] Das, A., Ringu, T., Ghosh, S., & Pramanik, N. (2023). A comprehensive review on recent advances in preparation, physicochemical characterization, and bioengineering applications of biopolymers. Polymer Bulletin, 80(7), 7247–7312.

[26] Gowthaman, N. S. K., Lim, H. N., Sreeraj, T. R., Amalraj, A., & Gopi, S. (2021). Advantages of biopolymers over synthetic polymers: social, economic, and environmental aspects. In Biopolymers and their industrial applications (pp. 351–372). Elsevier.

[27] Kukoyi, A. R. (2015). Economic impacts of natural polymers. In Natural Polymers: Industry Techniques and Applications (pp. 339–362). Cham: Springer International Publishing.

2 Mucilage and Gums: An Overview

2.1 Definition and Classification of Mucilage and Gums

Mucilage is a gel-like substance that consists of polysaccharides and is typically secreted by plants in response to environmental conditions like drought or injury, or to assist in seed germination (Figure 2.1) [1]. Mucilage plays a crucial role in plant hydration, storage of water, and protection against pathogens [1, 2]. It is usually water-soluble, but its structure allows it to retain moisture, making it an effective substance for hydrating and protecting plant tissues. Mucilage is often found in various parts of plants such as seeds, roots, and leaves [3].

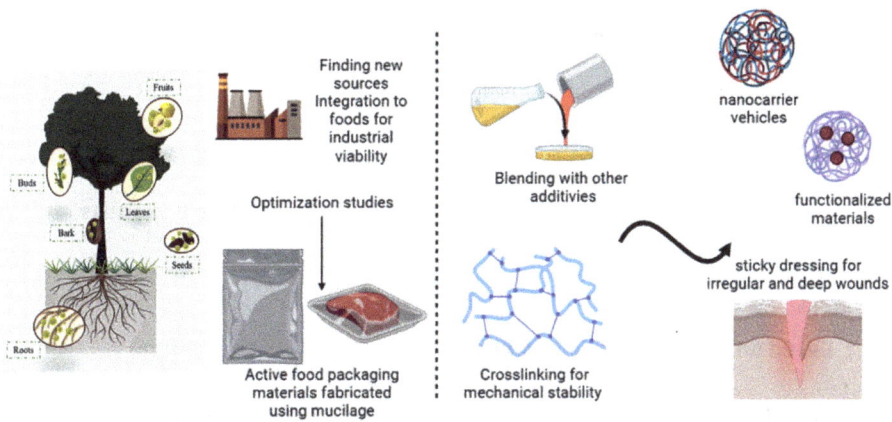

Figure 2.1: Mucilage. Reproduced with permission from [1].

Gums, on the other hand, are naturally occurring polysaccharides that have a more complex and highly branched structure compared to mucilage [1]. Gums are primarily exuded by plants in response to injury or stress and can act as a protective layer to seal wounds and prevent infection. They are usually water-soluble and can form viscous solutions or gels, depending on the conditions [2, 3]. Unlike mucilage, gums often have a more pronounced role in providing structural integrity to plants and may play a role in aiding the plant's defense mechanisms (Figure 2.2) [4].

Both mucilage and gums can be classified in various ways based on their chemical composition, botanical source, and physical properties. Their classification is typically divided into categories that help distinguish their structural characteristics and functions.

https://doi.org/10.1515/9783111673509-002

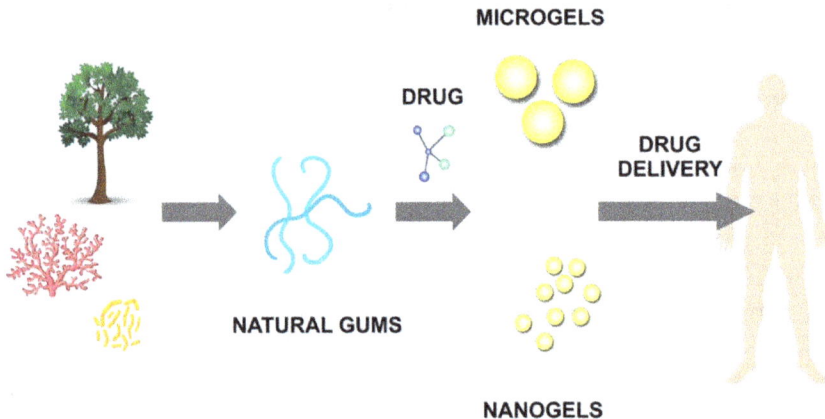

Figure 2.2: Gums. Reproduced with permission from [4].

2.2 Classification of Mucilage [5, 6]

Mucilage can be classified based on the type of polysaccharide it consists and its source. The key classifications are as follows.

2.2.1 Based on Polysaccharide Composition

– **Neutral Mucilages**: These are made up of simple, neutral sugars such as glucose, galactose, and mannose. Neutral mucilages do not contain acidic groups. Examples include **locust bean gum** and **guar gum**.
– **Acidic Mucilages**: These mucilages contain acidic polysaccharides, which often have sulfate or uronic acid groups. Examples of acidic mucilages include **pectin** and **xanthan gum**.
– **Mixed Mucilages**: These are composed of a mixture of neutral and acidic sugars and can also include other compounds like uronic acids. **Okra gum** is an example of mixed mucilage.

2.2.2 Based on Botanical Source [6]

Mucilage is produced by a variety of plant species, and its composition can vary depending on the source. Some of the most common plant sources include:
– **Seed Mucilages**: These mucilages are derived from the seeds of certain plants. They are commonly used as thickening agents and emulsifiers. Examples include **psyllium seed mucilage** and **flaxseed mucilage**.

- **Root Mucilages**: These are produced by the roots of certain plants, such as **okra** and **marshmallow** (*Althaea officinalis*). They are particularly valued for their high water-retention properties.
- **Leaf Mucilages**: Some plants, such as **aloe vera** and **hibiscus**, produce mucilage in their leaves, which have soothing and hydrating properties and are used in cosmetics and skin care products.

2.3 Gum Classification [5, 6]

Gums are classified based on their source and chemical structure. The following are the key classifications of gums.

2.3.1 Based on Botanical Source [6]

Gums are generally classified according to the plant from which they are derived. The major types of gums include:
- **Tree Gums**: These gums are exuded from trees, either from the bark or from specific structures like vessels or ducts. Common tree gums include **gum arabic** (from *Acacia* trees), **gum tragacanth** (from *Astragalus* species), and **gum karaya** (from *Sterculia* trees). Tree gums are usually water-soluble and are prized for their gelling and emulsifying properties.
- **Seed Gums**: Seed gums are obtained from the seeds of certain plants, such as **guar gum** and **locust bean gum**. These gums are often used as thickeners or stabilizers in food and pharmaceutical products.
- **Exudate Gums**: These are gums that are naturally exuded by plants when they are damaged or stressed. The exudates harden upon exposure to air, forming a gel-like substance. **Gum arabic** is one of the most famous examples of exudate gums.

2.3.2 Based on Chemical Composition [6]

Gums can also be classified according to their molecular structure and functional properties. The two major groups of gums based on chemical composition are:
- **Simple Gums**: These gums consist of a single type of sugar and are typically water-soluble. Examples of simple gums include **gum arabic** and **gum tragacanth**.
- **Complex Gums**: These gums are more complicated in structure and often consist of multiple sugar units. They tend to be less soluble in water compared to simple

gums but have better gelling properties. **Xanthan gum** and **guar gum** are examples of complex gums.

2.3.3 Based on Solubility and Functional Properties [5, 6]

- **Water-Soluble Gums**: These gums readily dissolve in water to form gels or viscous solutions. Common examples include guar gum, gum arabic, and xanthan gum. These gums are widely used as thickening agents and emulsifiers in the food industry.
- **Water-Insensitive Gums**: These gums are not soluble in water and instead form viscous gels upon mechanical action or heating. They are less commonly used but have applications in specialized industrial fields. Gum karaya and gum tragacanth are examples of water-insensitive gums.

2.4 Sources of Mucilage and Gums (Plants, Algae, and Microbial) [7]

Mucilages and gums are vital biopolymers found in a wide range of natural sources. These polysaccharides are exuded by plants, algae, and microorganisms and have essential biological functions in the organisms that produce them.

2.4.1 Plant-Based Sources of Mucilage and Gums [2, 7]

Plants are the primary producers of mucilage and gums. These polysaccharides are mainly exuded by specific plant tissues and serve functions such as water retention, defense, and seed dispersion. Some plant species are particularly renowned for their mucilage and gum content. The following are notable plant-based sources.

2.4.1.1 Mucilage from Plants [2]
Mucilage is commonly found in the seeds, roots, and stems of various plants. It serves as a protective mechanism, aiding in water retention and seed germination. Some of the major plant sources of mucilage include:
- **Flaxseed (*Linum usitatissimum*)**: Flaxseed mucilage is a water-soluble polysaccharide that is rich in galacturonic acid and glucose. It is primarily used for its soothing properties in food, pharmaceuticals, and cosmetic applications.
- **Okra (*Abelmoschus esculentus*)**: Okra mucilage is a natural gel-like substance composed of polysaccharides such as rhamnogalacturonan and arabinose. Okra

mucilage is used as a thickening agent in food products and has medicinal uses for treating ulcers and digestive issues.
- **Psyllium (*Plantago ovata*)**: Psyllium seed mucilage is a high-fiber material, primarily used as a bulk-forming laxative in pharmaceutical formulations. Its high-water retention makes it valuable in treating constipation and aiding in digestive health.
- **Aloe Vera (*Aloe barbadensis miller*)**: Aloe vera gel is a mucilaginous substance found in the inner leaf tissue of the plant. It has widespread use in skin care products, pharmaceuticals, and as a hydrating agent in food and beverages [8].

2.4.1.2 Gums from Plants [7]
Plant gums are exudates produced by certain plants when they are damaged. These exudates are often highly soluble in water and form viscous solutions or gels. Some major plant gums include:
- **Gum Arabic (*Acacia senegal*)**: One of the most widely used plant gums, gum arabic, is exuded from the Acacia tree. It is a complex mixture of polysaccharides and glycoproteins, widely used in food and beverages as an emulsifier, stabilizer, and thickening agent. It is also used in pharmaceuticals, cosmetics, and printing.
- **Guar Gum (*Cyamopsis tetragonoloba*)**: Guar gum is derived from the seeds of the guar plant and is a galactomannan polysaccharide. It is commonly used in the food industry as a thickening and stabilizing agent, as well as in hydraulic fracturing in the petroleum industry.
- **Gum Tragacanth (*Astragalus gummifer*)**: Gum tragacanth is a plant gum that is primarily sourced from the sap of Astragalus species. It is used as a stabilizer and emulsifier in the food industry, as well as in pharmaceutical and cosmetic formulations.
- **Gum Karaya (*Sterculia urens*)**: Gum karaya is exuded from the bark of the Sterculia tree and is commonly used in the food industry for its gelling properties. It is also employed in medical products such as adhesive plasters.

2.4.1.3 Algal Sources of Mucilage and Gums [8]
Algae, both marine and freshwater, are another important source of mucilages and gums. These biopolymers are primarily used in food, cosmetics, and industrial applications. Algal gums and mucilages typically consist of polysaccharides such as alginates, agar, and carrageenan, which have excellent gelling, thickening, and stabilizing properties:
- **Agar-Agar (*Gelidium* and *Gracilaria* species)**: Agar is a polysaccharide extracted from the cell walls of red algae. It forms gels when dissolved in boiling water and is used in microbiological research as a solidifying agent, as well as in food products like jellies, desserts, and ice cream.

- **Carrageenan (*Chondrus crispus*)**: Carrageenan is a gum obtained from red seaweed, particularly **Irish Moss (*Chondrus crispus*)**. It is widely used as a thickening, gelling, and stabilizing agent in dairy products, processed meats, and non-dairy beverages.
- **Alginates (*Macrocystis pyrifera, Laminaria* species)**: Alginates are extracted from brown algae and are composed of polysaccharides such as mannuronic acid and guluronic acid. Alginates are used in food processing, pharmaceuticals, and the textile industry as thickening agents and stabilizers.

2.4.1.4 Microbial Sources of Mucilage and Gums [9]

Microbial gums and mucilages are exopolysaccharides produced by various microorganisms such as bacteria, fungi, and yeasts. These gums are often used in industrial applications due to their ability to form highly viscous solutions and gels under specific conditions. Microbial gums are particularly significant in biotechnology and food industries.

- **Xanthan Gum (*Xanthomonas campestris*)**: Xanthan gum is a polysaccharide produced by the bacterium *Xanthomonas campestris*. It is widely used as a thickening, stabilizing, and emulsifying agent in food, cosmetics, and pharmaceuticals. Xanthan gum is highly effective at low concentrations and is used in products such as salad dressings, sauces, and toothpaste.
- **Gellan Gum (*Sphingomonas paucimobilis*)**: Gellan gum is produced by the bacterium *Sphingomonas paucimobilis* and is used in the food industry as a gelling agent, particularly in gel-like candies, jellies, and confectionery products. It is also used in biomedical applications for controlled drug release systems.
- **Dextran (*Leuconostoc mesenteroides*)**: Dextran is a polysaccharide synthesized by the bacterium *Leuconostoc mesenteroides*. It is used in medical applications, such as in blood volume expanders, as well as in food processing as a thickening agent.

2.5 Historical Uses and Significance [2, 7, 10]

The use of mucilage and gums can be traced back to ancient civilizations. For instance, evidence suggests that the earliest uses of these substances by humans may have been for medicinal purposes, as early humans observed the effects of natural substances on wounds and other ailments. The Egyptians, for example, are known to have used gums such as gum arabic (derived from the acacia tree) as a binding agent in their medicines and as a base for paints and inks. The earliest recorded use of gum arabic dates back to 1500 BCE, when it was mentioned in ancient Egyptian texts. Gum arabic's adhesive properties made it ideal for crafting a variety of products, including papyrus and other writing materials.

Mucilage, which is a sticky, gel-like substance, has been utilized for a long time, though it wasn't always understood in scientific terms. It was likely recognized for its thickening and water-retaining properties, which would have been of significant value in food preservation, textile production, and medicine. The ancient Greeks and Romans are believed to have used mucilage in various forms. One of the most notable examples is the use of mucilage from the flax plant, which was prized in the production of linen fabrics. The mucilage from flax seeds could be used as a natural adhesive or to improve the quality of textile fibers.

2.5.1 The Middle Ages and Renaissance (500 CE to 1500 CE)

During the Middle Ages and the Renaissance, knowledge of mucilages and gums expanded, albeit slowly. Arab traders were instrumental in the spread of gums and mucilages throughout Europe. For example, gum arabic became increasingly popular during this period, particularly in the formulation of inks for manuscript writing. It was widely recognized for its ability to improve the quality of ink by making it more stable and easier to apply.

In addition to its use in writing materials, gum arabic found applications in the food and beverage industries. It was used as a stabilizing agent in various forms of confectionery and beverages. Medieval alchemists and early scientists, like Paracelsus (1493–1541), explored the medicinal uses of these substances, although they lacked the modern understanding of their chemical properties.

The importance of mucilage in the textile industry continued to grow during this time as well. Flax fibers were increasingly treated with mucilage to make them easier to spin into thread and weave into fabric. Mucilage was also used in the production of certain medicines. For instance, mucilage from plant species such as marshmallow (*Althaea officinalis*) was applied as a remedy for coughs and sore throats.

The Age of Exploration, beginning in the late fifteenth century, led to increased global trade and the introduction of new plant species and gums. In particular, gum tragacanth, derived from the shrub *Astragalus* species, found in the Middle East and Asia, became highly valued for its thickening, emulsifying, and stabilizing properties. European botanists and apothecaries began studying these natural substances with increasing sophistication. By the seventeenth century, gum tragacanth was being used not only in food products but also in the creation of medicinal syrups and emulsions.

During the Enlightenment, with its emphasis on rationalism and scientific inquiry, there was a significant shift in understanding the chemical nature of natural substances, including gums and mucilage. The works of early chemists like Antoine Lavoisier in the late eighteenth century, who helped establish the principles of modern chemistry, eventually laid the foundation for a more systematic study of the molecular structures of these substances. As part of this process, researchers began to isolate and better understand the polysaccharide components of mucilage and gum, noting their unique properties, such as water retention and gel formation.

With the onset of the Industrial Revolution in the nineteenth century, mucilages and gums found expanded applications, particularly in the growing food, pharmaceutical, and textile industries. Advances in manufacturing and the development of new chemical processes allowed for the mass production and standardization of these substances. Gum arabic, gum tragacanth, and other natural gums became staples in industries, ranging from food production to cosmetics [11].

In food processing, gums were used as stabilizers, thickeners, and emulsifiers. For example, gum arabic became an important ingredient in the production of soft drinks, confectionery, and ice cream. Its ability to stabilize emulsions and improve texture made it ideal for these applications. Similarly, mucilage from plants such as guar and locust bean gum became widely used as thickening agents in soups, sauces, and salad dressings.

In the pharmaceutical industry, mucilage and gums continued to be important for their soothing, anti-inflammatory, and water-retaining properties. Plant-based mucilages, like that of the slippery elm (*Ulmus rubra*), were used in medicinal preparations for treating digestive ailments and respiratory conditions. They were also used in the formulation of capsules and tablets to aid in the delivery of active ingredients.

Furthermore, the textile industry continued to rely on mucilage for processes such as sizing and finishing fabrics. This was particularly important in the production of high-quality linen and cotton textiles, where mucilage was used to strengthen fibers and improve their durability.

By the twentieth century, the widespread industrialization of food production, pharmaceuticals, and textiles led to a greater reliance on mucilage and gums in a variety of commercial products. The discovery of synthetic polymers and the growing demand for natural alternatives gave rise to increased interest in natural gums and mucilages, which were viewed as eco-friendly and biodegradable. For example, the use of guar gum, derived from the seeds of the guar plant (*Cyamopsis tetragonoloba*), skyrocketed in the 1960s and 1970s, especially in the food industry. Guar gum became a popular thickening and stabilizing agent in everything from dairy products to gluten-free foods [12].

In addition to food and pharmaceutical uses, mucilage and gums began to find applications in newer industries such as biotechnology, agriculture, and cosmetics. In biotechnology, these substances were used in the production of biofilms and as carriers for controlled drug delivery. In agriculture, mucilage was used as a natural pesticide and as an agent to help plants retain moisture. In cosmetics, mucilage and gums became key ingredients in lotions, shampoos, and creams due to their hydrating and emulsifying properties.

The demand for natural gums continued to grow as consumer awareness about sustainability and the environment increased. Mucilage from various plant sources, such as chia seeds and flax seeds, gained popularity in health and wellness products due to their high fiber content and potential health benefits. These gums and mucilages are now frequently used as thickening agents in foods like smoothies, yogurt, and nutritional supplements [13].

2.6 Biological Significance of Mucilage and Gums

In nature, mucilages and gums serve several vital biological functions. Both are typically composed of complex sugars, which make them highly effective at retaining water, aiding in seed dispersal, and protecting the organism from environmental stressors such as drought, extreme temperatures, and mechanical damage:

- **Water Retention:** Mucilage is primarily water-soluble and has hygroscopic properties, meaning it can absorb and retain moisture. This function is crucial for many plants, especially in arid regions, where mucilage helps retain water within seeds, leaves, and stems, ensuring survival during periods of drought. For example, desert plants like *Cactus* and *Aloe vera* use mucilage to store water in their tissues.
- **Protection and Defense:** Mucilage and gums also serve as protective barriers. The sticky, viscous nature of these substances helps protect plant cells from dehydration, herbivores, and environmental threats. Some gums act as a physical barrier, preventing fungal and bacterial infections by forming protective layers over plant surfaces. For instance, the gum produced by the *Acacia* tree helps shield the plant from pathogens.
- **Seed Dispersal and Germination:** Mucilage plays an important role in seed dispersal and germination. In certain plants, the mucilage surrounding seeds aids in sticking to animals or water sources, ensuring the seeds travel to new locations. Additionally, mucilage can facilitate the absorption of water during seed germination, aiding in successful sprouting.

2.6.1 Culinary and Food Industry Applications

Mucilages and gums have a profound impact on the food industry, where they are primarily used for their ability to act as stabilizers, thickeners, and emulsifiers. Natural gums have a wide range of culinary applications, ranging from improving texture to enhancing the shelf life of food products:

- **Textural Enhancements:** Many gums, such as guar gum and xanthan gum, are used to improve the texture of food products. They are common ingredients in sauces, dressings, and gravies, where they help maintain consistency and prevent separation. In ice cream, gums prevent the formation of large ice crystals, giving the product a smooth texture [14].

- **Gluten-Free Alternatives:** Gums play an essential role in gluten-free baking. Since gluten is responsible for providing structure and elasticity to bread and other baked goods, gum-based ingredients like xanthan gum or guar gum are used as substitutes to help replicate the texture and mouthfeel typically associated with gluten. This makes them vital for creating gluten-free products, allowing individuals with celiac disease or gluten intolerance to enjoy foods that mimic traditional textures [15].
- **Stabilizing Properties:** In beverages like soft drinks and fruit juices, mucilages and gums help maintain uniformity by stabilizing the mixture. For example, gum arabic is commonly used in carbonated soft drinks to prevent the ingredients from separating and to maintain the beverage's quality over time [16].

2.7 Pharmaceutical and Medicinal Uses

Mucilage and gums are highly valued in the pharmaceutical industry for their therapeutic properties, which have been harnessed for centuries. Many plant-based gums possess anti-inflammatory, emollient, and soothing qualities, making them ideal for various medicinal applications:

- **Soothing and Anti-inflammatory Effects:** Mucilage has long been used in herbal medicine to soothe and heal mucous membranes. For example, slippery elm (*Ulmus rubra*) and marshmallow root (*Althaea officinalis*) contain mucilage, which is traditionally used to treat sore throats, coughs, and digestive discomfort. The mucilage forms a protective layer over mucous membranes, reducing irritation and promoting healing [17].
- **Laxative and Digestive Aid:** Mucilage is also commonly used as a mild, bulk-forming laxative. Psyllium husk, which contains a significant amount of mucilage, is widely known for its ability to promote regular bowel movements by absorbing water and adding bulk to stool. In addition to its laxative effect, mucilage from certain plants aids in soothing digestive ailments like acid reflux and ulcers by providing a protective coating for the stomach lining [18].
- **Controlled Drug Release:** Mucilage and gums are important in the pharmaceutical industry because they can be used as carriers for controlled drug release. Gums like xanthan and guar gum are used in the creation of controlled-release formulations, which allow drugs to be gradually released into the body over time. This has significant applications in treating chronic conditions, where consistent drug delivery is essential [19].

2.8 Industrial Applications

Beyond food and medicine, mucilages and gums have found vital roles in several industries, including textiles, cosmetics, and biotechnologies. Their ability to form films, bind particles, and improve texture has made them indispensable in the manufacturing of a wide array of products:

-   **Textile Industry:** Mucilage has historically been used in the textile industry as a sizing agent to coat fibers such as cotton and flax, making them easier to spin and weave. The sticky consistency of mucilage helps strengthen the threads and improves the fabric's smoothness and durability. The practice continues today, where mucilage is used in the finishing of textiles, particularly for specialty fabrics like linen [18].
-   **Cosmetic and Personal Care Products:** Gums and mucilage have emollient properties, which make them ideal ingredients in personal care products such as lotions, shampoos, and creams. Gums like guar gum are used as thickeners in shampoos and conditioners, giving them the desired consistency and allowing them to coat the hair for easier application. Similarly, in skincare products, mucilage acts as a moisturizing agent that hydrates and softens the skin [20].
-   **Paper and Paint Production:** Gum arabic has long been used as a binder in the production of inks and paints. Its adhesive properties allow it to be incorporated into various forms of media, from oil paints to watercolors. Additionally, it is used in the paper industry to coat papers, enhancing their smoothness and printability. This made gum arabic a key ingredient in the production of high-quality manuscripts during ancient times, and it continues to be used in printing and writing today [21].

2.9 Environmental and Sustainability Aspects

In an era of increasing concern for sustainability and environmental impact, mucilages and gums are gaining attention as eco-friendly alternatives to synthetic polymers. Their biodegradability and nontoxic nature make them ideal candidates for use in environmentally conscious products. Some gums, like guar gum and xanthan gum, are derived from renewable plant sources and can be produced with minimal environmental impact:

-   **Biodegradable Packaging:** With the rise of concerns about plastic waste, there is growing interest in developing biodegradable packaging materials using mucilage-based biopolymers. For example, researchers are exploring the use of mucilage from plants like okra and chia seeds to develop eco-friendly packaging alternatives. These materials can break down naturally, reducing reliance on plastic and contributing to environmental sustainability [22].

– **Water Conservation in Agriculture:** Mucilage also has a promising future in agriculture, particularly in soil water retention. Certain plant-based gums can be incorporated into soil to improve its ability to hold moisture, which is especially valuable in regions facing water scarcity. For example, mucilage from seeds like chia and fenugreek can help retain water around crops, reducing the need for frequent irrigation [23].

References

[1]   Goksen, G., Demir, D., Dhama, K., Kumar, M., Shao, P., Xie, F., . . . & Lorenzo, J. M. (2023). Mucilage polysaccharide as a plant secretion: Potential trends in food and biomedical applications. International Journal of Biological Macromolecules, 230, 123146.

[2]   Tosif, M. M., Najda, A., Bains, A., Kaushik, R., Dhull, S. B., Chawla, P., & Walasek-Janusz, M. (2021). A comprehensive review on plant-derived mucilage: characterization, functional properties, applications, and its utilization for nanocarrier fabrication. *Polymers, 13*(7), 1066.

[3]   Stephen, A. M., & Churms, S. C. (1995). Gums and mucilages. *FOOD SCIENCE AND TECHNOLOGY-NEW YORK-MARCEL DEKKER-*, 377–377.

[4]   Froelich, A., Jakubowska, E., Jadach, B., Gadziński, P., & Osmałek, T. (2023). Natural gums in drug-loaded micro-and nanogels. Pharmaceutics, 15(3), 759.

[5]   Glicksman, M. (2020). Origins and classification of hydrocolloids. In *Food hydrocolloids* (pp. 3–18). Crc Press.

[6]   De Barros, T. C., Leite, V. G., Pedersoli, G. D., Leme, F. M., Marinho, C. R., & Teixeira, S. P. (2023). Mucilage cells in the flower of Rosales species: reflections on morphological diversity, classification, and functions. *Protoplasma, 260*(4), 1135–1147.

[7]   Malabadi, R. B., Kolkar, K. P., & Chalannavar, R. K. (2021). Natural plant gum exudates and mucilage: pharmaceutical updates. *Int J Innov Sci Res Rev, 3*(10), 1897–1912.

[8]   Boney, A. D. (1981). Mucilage: the ubiquitous algal attribute. *British Phycological Journal, 16*(2), 115–132.

[9]   Haruna, S., Aliyu, B. S., & Bala, A. (2016). Plant gum exudates (Karau) and mucilages, their biological sources, properties, uses and potential applications: A review. *Bayero Journal of Pure and Applied Sciences, 9*(2), 159–165.

[10] Singh, R., & Barreca, D. (2020). Analysis of gums and mucilages. In *Recent Advances in Natural Products Analysis* (pp. 663–676). Elsevier.

[11] Jones, J. K. N., & Smith, F. (1949). Plant gums and mucilages. In *Advances in carbohydrate chemistry* (Vol. 4, pp. 243–291). Academic Press.

[12] Shiam, M. A. H., Islam, M. S., Ahmad, I., & Haque, S. S. (2025). A review of plant-derived gums and mucilages: Structural chemistry, film forming properties and application. *Journal of Plastic Film & Sheeting, 41*(2), 195–237.

[13] Amiri, M. S., Mohammadzadeh, V., Yazdi, M. E. T., Barani, M., Rahdar, A., & Kyzas, G. Z. (2021). Plant-based gums and mucilages applications in pharmacology and nanomedicine: a review. *Molecules, 26*(6), 1770.

[14] Ribes, S., Grau, R., & Talens, P. (2022). Use of chia seed mucilage as a texturing agent: Effect on instrumental and sensory properties of texture-modified soups. *Food Hydrocolloids, 123*, 107171.

[15] Dick, M., Limberger, C., Thys, R. C. S., de Oliveira Rios, A., & Flôres, S. H. (2020). Mucilage and cladode flour from cactus (Opuntia monacantha) as alternative ingredients in gluten-free crackers. *Food Chemistry, 314*, 126178.

[16] Soukoulis, C., Gaiani, C., & Hoffmann, L. (2018). Plant seed mucilage as emerging biopolymer in food industry applications. *Current Opinion in Food Science*, *22*, 28–42.

[17] Bahadur, S., Sahu, U. K., Sahu, D., Sahu, G., & Roy, A. (2017). Review on natural gums and mucilage and their application as excipient. *Journal of applied pharmaceutical research*, *5*(4), 13–21.

[18] Sharma, D. R., Sharma, A., Kaundal, A., & Rai, P. K. (2016). Herbal gums and mucilage as excipients for Pharmaceutical Products. *Research Journal of Pharmacognosy and Phytochemistry*, *8*(3), 145–152.

[19] Malabadi, R. B., Kolkar, K. P., & Chalannavar, R. K. (2021). Natural plant gum exudates and mucilage: pharmaceutical updates. *Int J Innov Sci Res Rev*, *3*(10), 1897–1912.

[20] Prajapati, V., Desai, S., Gandhi, S., & Roy, S. (2022). Pharmaceutical applications of various natural gums and mucilages. In *Gums, resins and latexes of plant origin: Chemistry, biological activities and uses* (pp. 25–57). Cham: Springer International Publishing.

[21] Jani, G. K., Shah, D. P., Prajapati, V. D., & Jain, V. C. (2009). Gums and mucilages: versatile excipients for pharmaceutical formulations. *Asian J Pharm Sci*, *4*(5), 309–323.

[22] Bhosale, R. R., Osmani, R. A. M., & Moin, A. (2014). Natural gums and mucilages: a review on multifaceted excipients in pharmaceutical science and research. *International Journal of Pharmacognosy and Phytochemical Research*, *15*(6), 4.

[23] Wadhwa, J., Nair, A., & Kumria, R. (2013). Potential of plant mucilages in pharmaceuticals and therapy. *Current drug delivery*, *10*(2), 198–207.

3 Sustainable Applications of Natural Polymers

3.1 Sustainable Applications of Natural Polymers

Natural polymers, derived from renewable biological sources, such as plants, animals, and microorganisms, offer significant advantages in promoting sustainability across various sectors. These biopolymers, including cellulose, starch, chitosan, alginate, and proteins like gelatin and silk, are biodegradable, nontoxic, and often biocompatible, making them ideal for eco-friendly applications [1]. In packaging, natural polymers are increasingly replacing conventional plastics due to their ability to decompose without leaving harmful residues, thus reducing environmental pollution [2, 3]. In agriculture, they are used as biodegradable films, seed coatings, and controlled-release fertilizers, enhancing crop yield while minimizing chemical runoff [4]. The biomedical field also benefits from their use in wound dressings, drug delivery systems, and tissue engineering, as they closely mimic natural tissue environments and degrade safely in the body [5]. Additionally, natural polymers play a crucial role in water purification, where they serve as bio-flocculants to remove contaminants without introducing harmful chemicals [6]. Their use in textiles, cosmetics, and construction materials further highlights their versatility and low environmental impact. As the world moves toward a circular economy, natural polymers present a sustainable alternative to synthetic materials, aligning with global efforts to reduce carbon footprints, limit resource depletion, and support a greener future.

3.2 Biodegradability and Environmental Benefits

In the modern world, the environmental impacts of synthetic materials, particularly plastics, have become a growing concern. The overuse of petroleum-based plastics and the accompanying accumulation of plastic waste in landfills and oceans have sparked a global environmental crisis.

3.2.1 Biodegradable Packaging

The global demand for biodegradable packaging has grown significantly as consumers and businesses seek alternatives to traditional plastics that are nonbiodegradable and persist in the environment for hundreds of years. Natural polymers such as starch, chitosan, cellulose, and polylactic acid (PLA) are gaining traction in the packaging industry due to their biodegradability, renewable sourcing, and reduced environmental impact [7].

https://doi.org/10.1515/9783111673509-003

3.2.2 Starch-Based Packaging

Starch is one of the most widely used natural polymers for biodegradable packaging. It is abundant, nontoxic, and biodegradable. Films and coatings made from starch can be used for food packaging, disposable cutlery, and bubble wraps. Starch-based packaging breaks down into water, carbon dioxide, and organic matter, significantly reducing landfill waste [7].

Environmental Impact: Starch is a renewable resource, and its use reduces reliance on petroleum-based plastics. It also produces fewer carbon emissions during production compared to synthetic plastics.

3.2.2.1 Chitosan Films

Chitosan, derived from chitin found in shellfish, is a biodegradable, nontoxic polymer with antimicrobial properties (Figure 3.1). This makes it an excellent choice for food packaging, especially in applications that require preservation, such as fruit and vegetable packaging. Chitosan films are also compostable, providing an environmentally friendly option for the food industry [8].

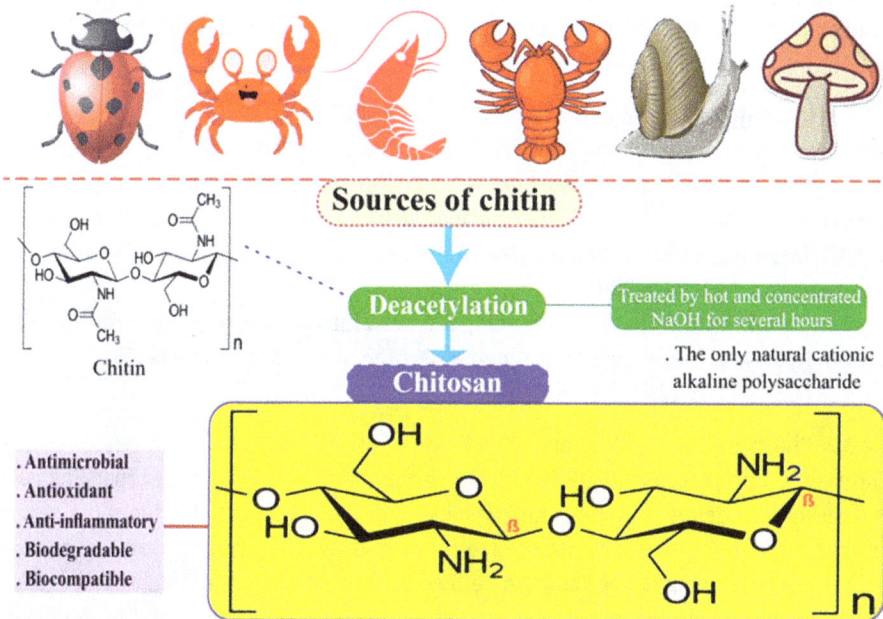

Figure 3.1: Chitosan. Reproduced with permission from [8].

Environmental Impact: Chitosan offers a sustainable alternative to plastic films by reducing waste and improving food preservation without the environmental hazards associated with synthetic plastic packaging.

3.2.2.2 Polylactic Acid (PLA)
PLA is made by fermenting plant sugars (usually derived from corn), making it a renewable and compostable polymer. It is commonly used for single-use packaging such as plastic cups, straws, and food containers. PLA can be composted in industrial composting facilities, where it breaks down within weeks, unlike petroleum-based plastics that persist for centuries [9].

    Environmental Impact: PLA reduces plastic waste in landfills, lowers carbon footprints, and lessens dependence on fossil fuels.

3.2.3 Medical and Pharmaceutical Applications

Natural polymers are increasingly being utilized in the medical and pharmaceutical fields due to their biocompatibility, biodegradability, and ability to reduce health risks. They are especially valuable in drug delivery systems, wound care, and surgical sutures, providing significant environmental benefits by reducing medical waste.

3.2.3.1 Hydrogels (e.g., Alginate, Agarose, and Cellulose)
Hydrogels are water-swollen polymers that are used extensively in drug delivery systems and wound healing. Alginate and agarose, both derived from seaweed, and cellulose-based hydrogels are used for controlled drug release, skin regeneration, and as wound dressings. These materials promote faster healing while being biodegradable, thus reducing medical waste.

    Environmental Impact: By reducing the need for nonbiodegradable alternatives, natural hydrogels lower the environmental burden of medical products [10].

3.2.3.2 Chitosan in Drug Delivery
Chitosan is a polysaccharide derived from chitin, found in the shells of crustaceans. It is used in drug delivery systems and tissue engineering applications due to its biodegradability and ability to promote cell growth. In drug delivery, chitosan can control the release of medicines, reducing side effects and improving treatment efficacy [11].

    Environmental Impact: Chitosan reduces the reliance on synthetic materials and ensures that the degradation products are nontoxic to the environment.

3.2.3.3 Silk Fibroin in Sutures

Silk fibroin, a protein derived from silkworms, is used to create sutures that degrade naturally within the body over time, eliminating the need for removal surgery. These sutures provide a safer and more convenient solution for patients while avoiding the disposal issues associated with synthetic sutures [12].

Environmental Impact: Silk-based sutures are biodegradable, reducing the accumulation of medical waste and supporting sustainable healthcare practices.

3.2.4 Textiles and Apparel

Natural polymers have long been used in textiles, with innovations continuing to emerge for creating more sustainable fabrics. Cotton, wool, and silk are traditional examples of natural fibers, while newer biopolymers like PLA and cellulose are offering additional sustainable options [13].

3.2.4.1 Cellulose-Based Fibers (e.g., Lyocell and Tencel)

Cellulose is a polysaccharide found in plant cell walls, and fibers such as Lyocell and Tencel are made from sustainably sourced wood pulp. These fibers are biodegradable and manufactured using a closed-loop process that recycles water and chemicals, minimizing environmental impact. Lyocell and Tencel are widely used in eco-friendly clothing lines due to their softness and sustainability [14].

Environmental Impact: Cellulose fibers require less water, energy, and chemicals compared to conventional cotton. They also decompose naturally at the end of their life cycle, reducing textile waste.

3.2.4.2 PLA Fabrics

PLA is a polymer derived from fermented plant sugars, and it is increasingly used in the production of biodegradable fabrics. PLA textiles, unlike traditional polyester, break down naturally, offering a viable alternative for sustainable fashion [15].

Environmental Impact: PLA fabrics are compostable and biodegradable, which helps reduce long-term textile waste and the reliance on fossil fuels.

3.2.5 Agriculture

In agriculture, natural polymers are being utilized for a variety of applications, such as improving soil health, enhancing crop growth, and reducing plastic pollution in farming environments.

3.2.5.1 Starch-Based Mulches

Starch-based biodegradable mulches are an excellent alternative to traditional plastic mulches used in agriculture. These mulches break down naturally in the soil, adding organic matter and improving soil quality without leaving harmful residues [3].

Environmental Impact: Starch-based mulches reduce plastic waste in agricultural fields and improve soil health by adding valuable organic matter.

3.2.5.2 Hydrogels in Agriculture

Natural polysaccharides, such as guar gum and xanthan gum, are used to create hydrogels that help improve water retention in soil. These hydrogels can absorb large quantities of water and release it slowly, reducing the need for frequent irrigation and ensuring better water usage in drought-prone areas [16].

Environmental Impact: By reducing water consumption and improving soil moisture retention, natural hydrogels help conserve water resources and support sustainable farming practices.

3.2.5.3 Biodegradable Plant Pots

Made from natural polymers like starch or cellulose, biodegradable plant pots are gaining popularity in gardening and agriculture. These pots break down in the soil, eliminating the need for plastic containers that would otherwise contribute to environmental pollution [17].

Environmental Impact: These pots reduce plastic waste in farming environments and enhance soil health by adding organic material.

3.2.6 Water Treatment

Natural polymers, particularly chitosan and pectin, are increasingly being used in water treatment to remove contaminants and purify water. These natural alternatives offer significant environmental advantages over synthetic chemicals [18].

3.2.6.1 Chitosan as a Flocculant

Chitosan can be used as a natural flocculant to remove heavy metals, oils, and dyes from wastewater. As a biodegradable material, it provides an eco-friendly alternative to synthetic flocculants, which can be harmful to the environment [19].

Environmental Impact: Chitosan's biodegradability ensures that it does not contribute to secondary pollution, making it a safer option for water treatment.

3.2.6.2 Alginate in Water Treatment

Alginate, derived from brown seaweed, is another natural polymer used in water treatment processes. Its ability to bind to particles and remove them from water makes it an effective natural coagulant [20].

Environmental Impact: Alginate is biodegradable and nontoxic, reducing the need for harmful chemical additives in water purification systems.

3.2.7 Construction and Building Materials

Research into the use of natural polymers in the construction industry is gaining momentum, particularly for biodegradable composite materials and insulation solutions [21].

3.2.7.1 Cellulose-Based Composites

Cellulose-based composites made from natural fibers and biodegradable resins are used in the production of sustainable building materials. These materials are used for insulation, cladding, and furniture [22].

Environmental Impact: These composites reduce carbon emissions associated with traditional building materials like cement and plastic. They also offer the benefit of being biodegradable at the end of their life cycle.

3.2.7.2 Mycelium-Based Materials

Mycelium, the root structure of fungi, is being explored as a sustainable alternative to traditional insulation, bricks, and even furniture. Mycelium-based products are biodegradable, nontoxic, and highly sustainable [23].

Environmental Impact: These materials provide an eco-friendly alternative to conventional building materials, reducing the environmental footprint of the construction industry.

3.2.8 Cosmetic and Personal Care Products

The cosmetic industry is increasingly turning to natural polymers for their emulsifying, thickening, and moisturizing properties. Many of these ingredients are biodegradable, offering a sustainable alternative to microplastics commonly found in synthetic personal care products [24].

3.2.8.1 Agar-Agar, Guar Gum, and Xanthan Gum

Agar-agar, guar gum, and xanthan gum are natural polysaccharides used in personal care products such as lotions, shampoos, and facial masks. These natural polymers help thicken and stabilize formulations, reducing the need for synthetic chemicals [24].

Environmental Impact: These natural alternatives reduce the environmental impact of personal care products, as they are biodegradable and come from renewable resources.

3.2.8.2 Chitosan in Cosmetics

Chitosan is used in cosmetics for its moisturizing, antibacterial, and film-forming properties. It is found in facial masks, skincare lotions, and other cosmetic products [25].

Environmental Impact: Chitosan's biodegradable nature helps reduce the use of synthetic polymers and microplastics, which can accumulate in aquatic ecosystems.

3.3 Challenges in Replacing Synthetic Polymers

Synthetic polymers have become an integral part of modern life, contributing to a variety of industries, from packaging and healthcare to automotive and electronics. One of the most significant challenges in replacing synthetic polymers is ensuring that alternative materials can match or exceed the performance characteristics of their synthetic counterparts. Synthetic polymers like polyethylene (PE), polypropylene (PP), and polystyrene (PS) are popular because of their excellent properties, including flexibility, high tensile strength, low cost, and ease of processing.

– **Polyethylene (PE):** Tensile strength = 20–30 MPa, elongation at break = 200–500%
– **Polypropylene (PP):** Tensile strength = 30–50 MPa, impact resistance = high
– **Polystyrene (PS):** Tensile strength = 45–70 MPa, glass transition temperature = 100 °C

For alternative materials, such as biopolymers (e.g., PLA, polyhydroxyalkanoates [PHA], and starch-based plastics), replicating the specific qualities of synthetic polymers is challenging. For instance, while PLA has gained attention due to its biodegradability and compostability, it falls short in terms of toughness and heat resistance when compared to traditional plastics like PE and PP. Challenges of heat resistance: Many biopolymers degrade at lower temperatures compared to synthetic polymers. PLA has a heat deflection temperature of 50–60 °C, which is much lower than that of PE (≈120 °C) and their mechanical properties: While some bio-based plastics have comparable strength, they often exhibit lower durability or impact resistance, which limits their applications in industries like automotive and construction.

3.3.1 Cost and Economic Viability

The economic feasibility of replacing synthetic polymers with bio-based or sustainable alternatives is a significant challenge. Bio-based polymers, while being eco-friendly, are often more expensive to produce due to the high cost of raw materials (e.g., corn, sugarcane, or agricultural waste); there are also complexities involved in processing these materials. The production cost of PLA is approximately **$2–4 per kilogram**, while conventional PE costs around **$0.80–1.50 per kilogram**. This price difference can make large-scale adoption of bio-based polymers impractical in many industries, particularly in sectors like packaging, where cost minimization is critical.

Additionally, the manufacturing processes for synthetic polymers are highly optimized and established, leading to economies of scale. Conversely, bio-based polymer production technologies are still in the development or scaling phase, leading to higher operational costs.

3.3.1.1 Key Cost-Related Challenges
- **Raw Material Costs**: Agricultural feedstocks can be subject to price volatility due to factors like weather conditions, market demand, and land use competition.
- **Manufacturing Infrastructure**: The infrastructure required to scale up the production of bio-based polymers is less developed than that for synthetic polymers, which increases investment and operational costs.

3.3.2 Limited Availability of Feedstock

While synthetic polymers are primarily derived from petroleum, which is widely available, the feedstocks for bio-based polymers (such as agricultural crops or waste materials) face limitations in terms of availability, scalability, and environmental impact.

For instance, many bio-based polymers rely on crops like corn, sugarcane, or potatoes. However, the use of crops for polymer production competes with food production, which raises concerns about food security. Moreover, large-scale agricultural operations often come with their own environmental concerns, such as water consumption, land use, and pesticide use:
- The global production of bio-based polymers is estimated to be around **6 million metric tons** annually, compared to the **360 million metric tons** of synthetic polymers produced each year. This stark contrast highlights the vast difference in scale between bio-based and synthetic polymer production.
- The land area required to produce bioplastics from crops is considerable. For example, producing **1 ton of PLA** may require approximately **3.3 hectares** of agricultural land.

As the demand for sustainable materials grows, securing a consistent and scalable supply of bio-based feedstocks becomes increasingly complex. Competing demands for land, food production, and bioplastics may limit the future availability of these resources.

3.3.3 Recycling and End-of-Life Management

Synthetic polymers like PE and PP are widely used in applications such as packaging, which typically end up in landfills or oceans due to their nonbiodegradability. Recycling these materials is challenging because of issues such as contamination, the complexity of sorting, and the limited infrastructure available for collection and processing.

Bioplastics, while biodegradable in certain conditions, also present challenges in terms of recycling. For example, PLA is compostable, but only under industrial composting conditions, not in typical landfill environments. Furthermore, PLA cannot be easily recycled with other plastics, complicating waste management systems [26]:

–   The global recycling rate for plastics is about **9%** of the total plastic waste produced, while **30–40%** of plastic waste is sent to landfills. The remaining plastic ends up in incinerators or as environmental pollution.
–   **PLA's biodegradation** rate is approximately **90% in industrial composting** over 45 days, but it can take much longer (several months to years) in natural environments like oceans or soil.

Additionally, the introduction of new polymers (bio-based or otherwise) could further complicate recycling systems. A shift toward sustainable alternatives requires upgrading or developing new recycling systems, which can be costly and logistically challenging.

3.3.4 Public Awareness and Acceptance

Another major hurdle in replacing synthetic polymers is public awareness and acceptance of alternative materials. Despite growing environmental awareness, the widespread use of synthetic polymers in daily life ranging from plastic bags to food containers – means that people are often reluctant to change their habits or embrace unfamiliar alternatives.

There is also skepticism about the performance and durability of new materials, especially if they are perceived to be inferior to traditional plastics. This challenge is compounded by a lack of clear labeling and standardized certifications that could help consumers make informed choices:

–   A study found that approximately **70% of consumers** globally are willing to pay
    a premium for sustainable products, but only **30%** actively seek out sustainable
    alternatives when shopping. This shows a gap between intention and action.

Consumer education is crucial to overcoming this challenge. As more alternatives
enter the market, educational campaigns will be essential to shift public perceptions
and behaviors.

3.3.5 Regulatory and Policy Challenges

Replacing synthetic polymers with bio-based alternatives also requires navigating a
complex regulatory landscape. Governments and international organizations have established various standards for the production, labeling, and disposal of plastics and
any new materials will need to meet these standards to ensure widespread adoption.
    Moreover, regulations regarding waste management, recycling, and the use of
biodegradable materials vary widely across regions, which can create confusion for
manufacturers and consumers alike. Inconsistent regulatory frameworks can hinder
the development and commercialization of bio-based polymers:
–   The European Union has set ambitious targets for plastic waste recycling, includ-
    ing a target of **50% recycling of plastic packaging by 2025**. This regulatory push
    is driving the development of sustainable alternatives but also highlights the bar-
    riers posed by existing regulations.

3.3.6 Environmental Trade-Offs

While bio-based polymers are often marketed as more sustainable than their synthetic counterparts, there are trade-offs that need to be carefully considered. For example, the environmental impact of growing the raw materials for bioplastics can be
significant, particularly in terms of water use, land degradation, and carbon emissions associated with agriculture and transportation:
–   The carbon footprint of PLA production is approximately **2–3 times** higher than
    that of conventional PE, primarily due to agricultural practices and transpor-
    tation.

Additionally, not all bio-based materials are biodegradable under all conditions. Some
may still contribute to microplastic pollution if they break down into small plastic
particles in marine environments.

3.3.7 Innovation and Technological Development

One of the most important factors in replacing synthetic polymers is the need for continued innovation in material science and technology. Bioplastics are still in the research and development phase for many applications, and scientists are exploring new sources, processing techniques, and blending technologies to overcome the limitations of bio-based polymers.

For instance, in the case of **PHA**, which are biodegradable plastics produced by bacteria, there is potential for improving their cost-effectiveness and scalability. Currently, PHAs are expensive to produce due to the high costs associated with bacterial fermentation and raw material inputs. However, with advancements in genetic engineering and fermentation processes, it is possible to make PHA production more cost competitive:

– The production cost of PHA currently stands at around **$4–6 per kilogram**, compared to **$1–2 per kilogram** for traditional petroleum-based plastics. While the price gap is narrowing, it remains a significant barrier.

Technological advancements in enzymatic processes, green chemistry, and renewable energy integration into polymer production could help reduce the environmental footprint and production costs of bio-based polymers. **Biocatalysts** and **enzymatic degradation techniques** are showing promise for improving the biodegradability and recycling of these new materials.

3.3.7.1 Key Technological Challenges

– **Scaling production**: Despite the growing interest in bioplastics, scaling up production to meet global demand remains a bottleneck. Large-scale manufacturing requires robust, efficient production systems that maintain cost competitiveness while minimizing environmental impact.
– **Material Optimization**: Many bio-based plastics still need improvements in properties such as **heat resistance**, **mechanical strength**, and **optical clarity**. These materials need to be further optimized to compete directly with traditional synthetic polymers, especially in high-performance applications.

3.3.8 Supply Chain Shifts and Adaptation

A significant shift in the global supply chain will be necessary to facilitate the mass adoption of alternative polymers. The entire process, from raw material procurement to manufacturing and waste disposal, must evolve. Currently, the supply chain for synthetic polymers is well-established and highly efficient. Transitioning to bio-based

materials could disrupt existing processes, leading to challenges in logistics, raw material sourcing, and manufacturing capabilities:
– **Global Plastic Production**: In 2020, 360 million metric tons of plastic were produced globally, with synthetic polymers dominating the market. Meanwhile, the production of bio-based plastics, although growing, still represented less than 5% of total plastic production globally.

A transition to bio-based plastics could result in the need for new facilities that can process plant-based feedstocks, such as crops or waste. Similarly, supply chain transparency would need to be enhanced to ensure the traceability and sustainability of bio-based polymer sources. Companies that currently rely on petrochemical-based supply chains may face logistical hurdles in sourcing bio-based raw materials that are geographically distant or subject to agricultural constraints.

3.3.8.1 Challenges in Supply Chain
– **Sourcing Raw Materials**: Bio-based polymer production depends on agricultural feedstocks that can be subject to global crop shortages, price volatility, and seasonality.
– **Infrastructure Overhaul**: Existing manufacturing facilities would need to be repurposed to accommodate new biopolymer production techniques, requiring significant investment and time.

3.3.9 Consumer Behavior and Lifestyle Changes

One of the major hurdles that could slow down the transition to sustainable polymers is consumer behavior. As mentioned previously, a large portion of consumers express an interest in eco-friendly products but are not always willing to make the necessary changes to their buying habits. Despite growing environmental concerns, many people are still hesitant to adopt alternatives to conventional plastics due to perceived inconveniences, such as higher costs, unfamiliar materials, or limited product choices.

The success of bioplastics in the consumer market will depend heavily on consumer education and behavioral change. It is essential to communicate the environmental benefits of switching to bio-based materials, not just from a recycling perspective but also in terms of the broader ecological impact:
– According to Nielsen's 2019 Global Sustainability Report, 73% of global consumers said they would change their consumption habits to reduce their environmental impact. However, the report also showed that only 34% of consumers actively follow through with this intention when it comes to their everyday purchases.

3.3.9.1 Overcoming Behavioral Barriers

– **Awareness Campaigns**: Governments and businesses need to ramp up education campaigns to inform the public about the environmental costs of conventional plastics and the benefits of switching to alternatives. Clear labeling systems that identify products as biodegradable or made from renewable resources could encourage consumers to make informed choices.
– **Consumer Incentives**: Financial incentives, such as tax breaks or subsidies for products made from renewable materials, could motivate consumers to support sustainable products. Additionally, businesses could offer discounts for customers who choose eco-friendly options, further driving the demand for bio-based plastics.

3.3.10 Interdisciplinary Collaboration

Another important consideration in overcoming the challenges of replacing synthetic polymers is fostering interdisciplinary collaboration. Replacing synthetic polymers requires expertise not only in polymer chemistry but also in agricultural science, environmental policy, industrial engineering, and consumer behavior. Working together, these fields can accelerate the development, adoption, and scaling of bio-based plastics.

For example, partnerships between scientists and industry stakeholders could lead to more efficient **plant-based feedstocks** for bio-polymers, which could lower production costs. Similarly, close cooperation with waste management organizations and municipalities could lead to new recycling systems specifically designed for bio-based materials:

– **Interdisciplinary Collaboration**: The **bio-based plastics market** is projected to reach a value of **$35 billion** by 2027, growing at a CAGR of approximately **20%**. This growth is being driven by collaborative efforts between scientists, engineers, and businesses in the field of material science and sustainability.

3.3.11 Policy and Global Standardization

For success in global transition to sustainable polymers, a unified approach to regulations and standards is essential. Currently, different countries have varying policies when it comes to the regulation of synthetic and bio-based plastics. Some regions, such as the European Union, are ahead in terms of creating regulations that promote biodegradable alternatives, while other regions have weaker policies or none at all.

The establishment of global standards for bio-based polymers would encourage widespread adoption by providing manufacturers with clear guidelines for production and labeling. Moreover, regulatory frameworks could incentivize the production of sustainable materials by providing tax breaks, subsidies, or preferential treatment in government procurement:

– **EU Plastics Strategy**: The European Union's **Circular Economy Action Plan**, part of its Green Deal, aims to reduce plastic waste and increase recycling rates to **50%** by 2025. This push is driving demand for bio-based alternatives, but much work remains in aligning with global regulations.

A harmonized international regulatory framework could significantly reduce market fragmentation, enabling a smoother transition from synthetic to sustainable polymers.

3.4 Current Trends in Sustainable Materials Research

3.4.1 Biopolymers and Bio-based Plastics

– **Focus on Biodegradability**
One of the most pressing issues related to synthetic polymers, such as PE and PS, is their **nonbiodegradability**. Researchers are working to develop bio-based plastics that degrade more rapidly and safely in natural environments. **PLA** and **PHA** are two notable biopolymers that have shown promise in replacing conventional plastics:
  – **PLA**: Derived from renewable resources like corn and sugarcane, PLA is a biodegradable thermoplastic polymer. Researchers are focused on improving PLA's mechanical properties (like impact resistance) to make it suitable for a wider range of applications.
  – **PHA**: Produced by bacteria through fermentation processes, PHAs are another biodegradable polymer that can be used in packaging and medical applications. Recent advancements in fermentation technology aim to reduce the production cost of PHA, making it more competitive with petroleum-based plastics.

3.4.1.1 Current Research
– Biodegradable polymers that perform well in real-world conditions (e.g., in oceans and soil) and offer better shelf life
– Development of **composite biopolymers** that combine bio-based plastics with natural fibers to enhance properties such as strength and heat resistance

3.4.2 Recycling and Circular Economy

3.4.2.1 Improved Recycling Technologies
As the global demand for plastics continues to rise, so does the need for efficient and scalable recycling technologies. Conventional recycling processes often fail to reclaim plastics with mixed compositions, resulting in the accumulation of waste in landfills or oceans. This has spurred research into improving both the mechanical and chemical recycling of polymers.

3.4.2.2 Trends
- **Chemical Recycling**: Also known as **feedstock recycling**, this process breaks down plastics into their monomers or other useful chemicals, allowing them to be reused to create new products. The aim is to recycle plastics that are otherwise difficult to process using mechanical methods. Companies are investing in pyrolysis and catalytic cracking technologies to scale chemical recycling.
- **Upcycling**: Upcycling involves turning waste plastics into higher-value products rather than just downcycling. Research is exploring how to use waste plastics to create materials with properties that can outperform virgin plastic products.
- **Recycling of Mixed Plastics**: Developing technologies to sort and process mixed polymer waste without requiring labor-intensive manual separation is a growing area of research. Innovations in sorting and **machine learning** are improving the efficiency of plastic recycling.

3.4.3 Carbon-Negative and Carbon-Capture Materials

3.4.3.1 Materials for Carbon Sequestration
With the urgency of addressing climate change, there is increasing interest in materials that can not only reduce carbon emissions but actually **capture and store carbon**. Researchers are looking at a variety of novel materials for carbon sequestration, ranging from natural substances to synthetic alternatives:
- **Carbon Nanomaterials**: Researchers are investigating **carbon nanotubes (CNTs)** and **graphene** for their ability to absorb and store CO_2. These materials have high surface areas, which makes them ideal for capturing and storing carbon.
- **Biochar**: Created from organic materials through pyrolysis, biochar is a form of charcoal that can be used as a soil amendment to increase carbon storage in the ground. The material also has applications in water filtration, energy storage, and as a building material.

3.4.3.2 Current Research
– Development of carbon-capturing **cement** and concrete materials that absorb CO_2 as they harden
– New strategies to make carbon sequestration a profitable activity by integrating it with agriculture and forestry practice.

3.4.4 Green Chemistry and Sustainable Synthesis

3.4.4.1 Nontoxic, Renewable Synthesis Routes
Green chemistry focuses on minimizing the environmental impact of chemical processes and materials. Researchers are exploring sustainable synthesis routes for materials using renewable resources while avoiding hazardous chemicals. Key developments in this area include:
– **Bio-based Synthesis**: Using **plant-derived feedstocks** or agricultural waste as raw materials for producing chemicals and materials. This reduces dependency on petrochemical-derived substances, which have a high environmental cost [27].
– **Solvent-Free Processes**: Traditional chemical manufacturing often relies on toxic solvents that are not only harmful to the environment but also expensive to dispose of. Sustainable materials research is focusing on solvent-free or **green solvents** like water, supercritical CO_2, or ionic liquids for the synthesis of materials [28].

3.4.4.2 Current Research
– Use of **enzymatic processes** in polymerization and other material production, which offer a non-toxic and energy-efficient alternative to conventional chemical reactions
– Innovations in the production of bio-based **epoxy resins** for use in coatings and composites, a widely used material in construction and electronics

3.4.5 Energy-Efficient and Smart Materials

3.4.5.1 Materials for Energy Efficiency
As part of the effort to reduce energy consumption, research into materials with **high thermal insulation, low energy demand**, and **smart functionalities** is booming. These materials can lead to more sustainable buildings, energy systems, and even wearable technology:
– **Thermoelectric Materials**: These materials convert heat directly into electricity and vice versa. They are being used in applications such as waste heat recovery

systems in industrial processes, buildings, and even wearable technology to power devices from body heat.

– **Phase-Change Materials (PCMs)**: PCMs are materials that absorb or release heat as they change phase (e.g., from solid to liquid). They are being researched for use in energy-efficient buildings, where they can help maintain consistent temperatures, reducing heating and cooling energy consumption.

– **Piezoelectric and Triboelectric Materials**: These materials generate electrical energy from mechanical stress (e.g., from walking or wind). They are being integrated into self-powered systems, such as **wearable electronics** and **smart textiles** [29].

3.4.5.2 Current Research

– Development of **advanced thermal insulation materials** for buildings, reducing heating and cooling needs by enhancing the natural flow of energy

– Energy-harvesting materials integrated into urban environments such as **smart pavements** or **wearable devices** that collect energy from motion or light

3.4.6 Sustainable Construction Materials

3.4.6.1 Green Cement and Alternative Aggregates

The construction industry is one of the largest contributors to global carbon emissions. Cement production alone accounts for about **8% of global CO_2 emissions**. As a result, significant research is focused on developing sustainable alternatives to traditional construction materials:

– **Geopolymer Cement**: This type of cement, made from industrial waste like fly ash or slag, has a lower carbon footprint than traditional Portland cement. Geopolymers can also exhibit better resistance to heat, corrosion, and chemical attack, making them ideal for certain high-performance applications.

– **Recycled Concrete Aggregates**: By using crushed concrete from demolished buildings as aggregate, the need for virgin sand and gravel is reduced, conserving natural resources and reducing environmental impact.

– **Hempcrete and Other Bio-based Building Materials**: **Hempcrete** is a composite material made from hemp stalks and lime, providing excellent insulation properties while absorbing carbon from the atmosphere. Researchers are also exploring other bio-based composites, such as **mycelium-based materials** (fungal-based) that grow into sustainable building blocks [30].

3.4.6.2 Current Research

– Development of **bio-cement** that can self-heal cracks by incorporating living bacteria into the mix, which could improve the durability of buildings and reduce maintenance needs
– Exploring **sustainable wood alternatives** and using **reclaimed wood** for construction to reduce deforestation

3.4.7 Water-Saving and Purification Materials

3.4.7.1 Water Purification and Desalination

In the context of global water scarcity, researchers are developing advanced materials for **water purification, filtration**, and **desalination** that are more efficient and cost-effective than traditional technologies:

– **Graphene-Based Filters**: Graphene oxide membranes have shown promise for desalination and water filtration, as they allow water molecules to pass through while blocking larger contaminants. These membranes are more energy-efficient than conventional reverse osmosis methods.
– **Solar Desalination**: Solar-powered desalination systems use **solar energy** to convert seawater into freshwater. **Photothermal materials**, which absorb sunlight and convert it into heat, are central to these technologies [31].

3.4.7.2 Current Research

– Developing **self-cleaning filters** for use in rural and remote areas, where access to fresh water is limited
– Advancing low-cost, scalable **membrane materials** for large-scale desalination plants to make freshwater more accessible in arid regions

3.4.8 Advanced Biomaterials and Green Composites

3.4.8.1 Natural Fiber-Reinforced Composites

One of the most promising areas in sustainable materials research is the development of **natural fiber-reinforced composites**. These composites combine renewable natural fibers such as **hemp, jute, flax**, and **bamboo** with biodegradable matrices (such as PLA, PHA, or even lignin) to replace traditional petroleum-based composites, such as fiberglass and carbon fiber:

– **Advantages**: Natural fibers are renewable, lightweight, and biodegradable, offering a sustainable alternative to synthetic fibers. Furthermore, these materials often have superior thermal insulation and acoustic properties compared to traditional composites.

–   **Research Focus**: Enhancing the strength, durability, and resistance to water ab-
    sorption of natural fiber composites without compromising their environmental
    benefits. Techniques such as surface treatment of fibers (e.g., **alkali treatment** or
    nanotechnology coatings) are being explored to improve the bonding between
    fibers and matrices [32, 33].

3.4.8.2 Current Research

–   **Hybrid Composites**: Combining natural fibers with other biodegradable or bio-
    based polymers (such as **(PLA** and **PHA)** to achieve desirable properties, such as
    improved mechanical strength, durability, and flexibility, while ensuring com-
    plete biodegradability after use
–   **Green Reinforcement Materials**: Using **cellulose nanocrystals (CNCs)** or **cellu-
    lose nanofibers** as reinforcements for bio-based composites, improving the stiff-
    ness and strength-to-weight ratio without increasing environmental impact

3.4.9 Nanomaterials for Sustainable Solutions

3.4.9.1 Nanotechnology in Material Sustainability

Nanomaterials, due to their small scale and unique properties, are enabling new opportunities for sustainable materials. These materials can enhance the performance
of existing products, improve energy efficiency, and offer innovative solutions to environmental challenges. Nanomaterials have applications across various industries,
from **construction** to **electronics** and **medicine**, contributing to more efficient use of
resources:

–   **Self-Cleaning Materials**: Researchers are exploring **nanocoatings** for surfaces
    that make them self-cleaning and hydrophobic. These surfaces reduce the need
    for chemical cleaners, thus decreasing environmental pollution.
–   **Nanomaterials for Energy Storage**: In the field of **energy storage**, nanomateri-
    als like **graphene** and **CNTs** are used to enhance the performance of batteries
    and supercapacitors, improving energy density, charge cycles, and overall effi-
    ciency. This is particularly important for **renewable energy systems**, such as
    solar and wind, which rely on efficient storage solutions.
–   **Sustainable Water Filtration**: **Nanofiltration membranes** are being developed
    for more efficient water purification. They have the potential to remove even the
    smallest particles from water, such as viruses and bacteria, with lower energy
    consumption compared to traditional filtration methods [34–36].

3.4.9.2 Current Research

–   **Nanocellulose-Based Materials**: The development of **CNCs** and **nanofibrils** de-
    rived from plant sources is gaining traction due to their sustainability. These ma-

terials are biodegradable, nontoxic, and show great potential as reinforcements in composites and for water filtration, as well as in packaging applications.

– **Green Synthesis of Nanomaterials**: Nanoparticles typically require energy-intensive or toxic chemical processes for synthesis. Researchers are exploring **green synthesis methods**, where nanomaterials are synthesized using plant extracts, microorganisms, or waste by-products, making the process more environmentally friendly.

3.4.10 Sustainable Textiles and Eco-friendly Fabrics

3.4.10.1 Circular Textiles and Recycled Fabrics

The fashion and textile industry is one of the most resource-intensive sectors in terms of water usage, energy consumption, and waste generation. In response to growing concerns over the environmental impact of textile production, there is increasing research into **circular textiles** materials that can be continuously recycled and up-cycled without degrading in quality:

– **Recycled Fabrics**: Research into the recycling of fabrics, particularly synthetic textiles like polyester and nylon, is a key trend. Techniques to recycle garments and textiles into **high-quality fabrics** have the potential to significantly reduce the need for virgin materials.

– **Eco-friendly and Biodegradable Fabrics**: Innovations in **bio-fabrics**, such as **fungal leather** (mycelium), **algae-based fabrics**, and **bamboo textiles**, are gaining attention due to their minimal environmental footprint. These materials offer biodegradable or compostable alternatives to traditional synthetic fibers, which often persist in landfills for decades.

– **Waterless Dyeing**: Traditional fabric dyeing processes are water-intensive and polluting. New **waterless dyeing** technologies, such as **supercritical CO_2 dyeing**, offer a more sustainable alternative by using carbon dioxide in its supercritical state to dye fabrics without the need for water or harmful chemicals [37, 38].

3.4.10.2 Current Research

– **Recycling Technology for Blended Fabrics**: Developing technologies to separate mixed fibers like polyester-cotton blends, which currently pose a significant challenge to recycling processes

– **Biodegradable Fibers for Clothing**: Research into biodegradable fibers such as **hemp**, **Tencel**, and **sisal, which** can be used for creating sustainable fashion products that break down naturally after disposal, thus reducing environmental pollution

3.4.11 Sustainable Electronics and Green Electronic Materials

3.4.11.1 Eco-friendly Semiconductors and Conductive Materials

The growing e-waste problem has prompted researchers to explore alternatives to the harmful metals and chemicals used in traditional electronics manufacturing. **Sustainable electronics** focus on reducing toxicity, increasing the longevity of devices, and making it easier to recycle electronic products at the end of their life cycle:

- **Bio-based Electronics**: Research into bio-based semiconductors made from materials like **organic polymers**, **biodegradable plastics**, or **biopolymers** is a burgeoning area. These materials can replace traditional silicon and metal-based components, making electronics less hazardous when disposed of.
- **E-Waste Recycling**: Efforts are underway to make it easier to recycle metals and rare earth elements used in electronics, like **gold**, **silver**, and **copper**. New **e-waste recycling technologies** aim to extract these valuable materials efficiently and in an environmentally responsible manner.
- **Printed and Flexible Electronics**: The development of **printed electronics** and **flexible displays** that use **conductive inks** made from **silver nanoparticles** or **graphene** offers a more sustainable alternative to conventional rigid electronics. These materials can be printed directly onto substrates like paper, textiles, or biodegradable plastics, reducing the need for traditional electronic fabrication processes [39–41].

3.4.11.2 Current Research

- **Biodegradable Circuits**: Research into developing biodegradable electronic circuits made from **biopolymers** or **organic semiconductors** to address the growing e-waste crisis
- **Energy-Efficient Displays**: Development of **energy-efficient OLEDs** (organic light-emitting diodes) and **micro-LED** technology to reduce energy consumption in display devices

3.4.12 Food Packaging and Sustainable Food Systems

3.4.12.1 Biodegradable and Edible Packaging

In the food industry, packaging is one of the largest sources of waste. **Single-use plastic packaging** has long been a source of environmental concern. As a result, researchers are exploring **biodegradable**, **compostable**, and even **edible packaging materials** that can reduce food waste and pollution:

- **Edible Packaging**: Derived from **seaweed**, **rice**, or **cornstarch**, edible packaging can be consumed along with food, reducing the need for waste disposal. These materials are also biodegradable, meaning they can break down without leaving harmful residues behind.

- **Plant-Based Plastics**: Packaging made from plant-derived materials such as **PLA**, **PHA**, and **starch-based plastics** are becoming increasingly popular. These materials degrade more quickly than petroleum-based plastics and have lower environmental footprints [42–44].

3.4.12.2 Current Research

- **Active Packaging**: Developing **active food packaging** that can extend shelf life by incorporating antimicrobial agents or natural preservatives. These packages help preserve food without relying on chemical preservatives or excessive plastic.
- **Innovative Edible Films**: Creating edible films for food packaging that have superior **barrier properties** (to moisture, oxygen, etc.) to maintain food quality while being safe to consume or biodegradable.

3.4.13 Smart Agriculture and Sustainable Agricultural Materials

3.4.13.1 Agricultural Waste as a Resource

A major challenge in sustainable materials research is the effective utilization of **agricultural waste**. Many crops and plants produce large amounts of waste that are often discarded or burned, contributing to pollution. However, agricultural waste such as **rice husks**, **corn stalks**, and **sugarcane bagasse** are increasingly being used as feedstocks for the production of sustainable materials:

- **Bio-based Insulation**: Agricultural by-products like **hemp** and **straw** are being used to make eco-friendly building insulation materials.
- **Bioplastics from Agricultural Waste**: Converting agricultural waste into bio-based plastics or composite materials that can be used in packaging or construction [45, 46].

3.4.13.2 Current Research

- **Upcycling Agricultural Waste** into high-value materials, such as **biocomposites** for construction and packaging, as well as **biofuels** for energy production
- Exploring new **biodegradable agricultural films** and **mulch materials** made from natural fibers to reduce plastic pollution in farming

References

[1]   Silva, A. C., Silvestre, A. J., Vilela, C., & Freire, C. S. (2021). Natural polymers-based materials: A contribution to a greener future. *Molecules, 27*(1), 94.
[2]   Muthukumaran, P., Suresh Babu, P., Karthikeyan, S., Kamaraj, M., & Aravind, J. (2021). Tailored natural polymers: a useful eco-friendly sustainable tool for the mitigation of emerging pollutants: a review. *International Journal of Environmental Science and Technology, 18*(8), 2491–2510.
[3]   Mironescu, M., Lazea-Stoyanova, A., Barbinta-Patrascu, M. E., Virchea, L. I., Rexhepi, D., Mathe, E., & Georgescu, C. (2021). Green design of novel starch-based packaging materials sustaining human and environmental health. *Polymers, 13*(8), 1190.
[4]   Siracusa, V., & DALLA ROSA, M. (2018). Sustainable Packaging, in "Sustainable Food Systems from Agriculture to Industry". In *Sustainable Food Systems from Agriculture to Industry* (Vol. 1, pp. 275–307). Academic Press.
[5]   Zhao, L., Zhou, Y., Zhang, J., Liang, H., Chen, X., & Tan, H. (2023). Natural polymer-based hydrogels: From polymer to biomedical applications. *Pharmaceutics, 15*(10), 2514.
[6]   Guo, H., Qin, Q., Chang, J. S., & Lee, D. J. (2023). Modified alginate materials for wastewater treatment: Application prospects. *Bioresource technology, 387*, 129639.
[7]   Shiam, M. A. H., Islam, M. S., Ahmad, I., & Haque, S. S. (2025). A review of plant-derived gums and mucilages: Structural chemistry, film forming properties and application. *Journal of Plastic Film & Sheeting, 41*(2), 195–237.
[8]   Yarahmadi, A., Dousti, B., Karami-Khorramabadi, M., & Afkhami, H. (2024). Materials based on biodegradable polymers chitosan/gelatin: a review of potential applications. Frontiers in Bioengineering and Biotechnology, 12, 1397668.
[9]   Oksman, K., Skrifvars, M., & Selin, J. F. (2003). Natural fibres as reinforcement in polylactic acid (PLA) composites. *Composites science and technology, 63*(9), 1317–1324.
[10] Malabadi, R. B., Kolkar, K. P., & Chalannavar, R. K. (2021). Natural plant gum exudates and mucilage: pharmaceutical updates. *Int J Innov Sci Res Rev, 3*(10), 1897–1912.
[11] Bernkop-Schnürch, A., & Dünnhaupt, S. (2012). Chitosan-based drug delivery systems. *European journal of pharmaceutics and biopharmaceutics, 81*(3), 463–469.
[12] Li, X., Luo, Y., Yang, F., Chu, G., Li, L., Diao, L., . . . & Lyu, G. (2023). In situ-formed micro silk fibroin composite sutures for pain management and anti-infection. *Composites Part B: Engineering, 260*, 110729.
[13] Bao, H., Hong, Y., Yan, T., Xie, X., & Zeng, X. (2024). A systematic review of biodegradable materials in the textile and apparel industry. *The Journal of The Textile Institute, 115*(7), 1173–1192.
[14] Bledzki, A. K., & Gassan, J. (1999). Composites reinforced with cellulose based fibres. *Progress in polymer science, 24*(2), 221–274.
[15] Farrington, D. W., Lunt, J., Davies, S., & Blackburn, R. S. (2008). Poly (lactic acid) fibers (PLA). *Polyesters and polyamides*, 140–170.
[16] Kaur, P., Agrawal, R., Pfeffer, F. M., Williams, R., & Bohidar, H. B. (2023). Hydrogels in agriculture: Prospects and challenges. *Journal of Polymers and the Environment, 31*(9), 3701–3718.
[17] Tomadoni, B., Merino, D., & Alvarez, V. A. (2020). Biodegradable materials for planting pots. *Adv. Appl. Bio-Degrad. Green Compos, 68*, 85.
[18] Resende, J. F., Paulino, I. M. R., Bergamasco, R., Vieira, M. F., & Vieira, A. M. S. (2023). Hydrogels produced from natural polymers: A review on its use and employment in water treatment. *Brazilian Journal of Chemical Engineering, 40*(1), 23–38.
[19] Zeng, D., Wu, J., & Kennedy, J. F. (2008). Application of a chitosan flocculant to water treatment. *Carbohydrate polymers, 71*(1), 135–139.
[20] Devrimci, H. A., Yuksel, A. M., & Sanin, F. D. (2012). Algal alginate: A potential coagulant for drinking water treatment. *Desalination, 299*, 16–21.

[21] Tamošaitienė, J., Parham, S., Sarvari, H., Chan, D. W., & Edwards, D. J. (2024). A review of the application of synthetic and natural polymers as construction and building materials for achieving sustainable construction. *Buildings, 14*(8), 2569.

[22] Dufresne, A. (2008). Cellulose-based composites and nanocomposites. In *Monomers, polymers and composites from renewable resources* (pp. 401–418). Elsevier.

[23] Alemu, D., Tafesse, M., & Mondal, A. K. (2022). Mycelium-based composite: the future sustainable biomaterial. *International journal of biomaterials, 2022*(1), 8401528.

[24] Alves, T. F., Morsink, M., Batain, F., Chaud, M. V., Almeida, T., Fernandes, D. A., . . . & Severino, P. (2020). Applications of natural, semi-synthetic, and synthetic polymers in cosmetic formulations. *Cosmetics, 7*(4), 75.

[25] Morganti, P., Morganti, G., & Coltelli, M. B. (2023). Natural polymers and cosmeceuticals for a healthy and circular life: the examples of chitin, chitosan, and lignin. *Cosmetics, 10*(2), 42.

[26] Sharma, D. R., Sharma, A., Kaundal, A., & Rai, P. K. (2016). Herbal gums and mucilage as excipients for Pharmaceutical Products. *Research Journal of Pharmacognosy and Phytochemistry, 8*(3), 145–152.

[27] Kar, S., Sanderson, H., Roy, K., Benfenati, E., & Leszczynski, J. (2021). Green chemistry in the synthesis of pharmaceuticals. *Chemical Reviews, 122*(3), 3637–3710.

[28] Ganesh, K. N., Zhang, D., Miller, S. J., Rossen, K., Chirik, P. J., Kozlowski, M. C., . . . & Voutchkova-Kostal, A. M. (2021). Green chemistry: a framework for a sustainable future. *ACS omega, 6*(25), 16254–16258.

[29] Ke, Y., Zhou, C., Zhou, Y., Wang, S., Chan, S. H., & Long, Y. (2018). Emerging thermal-responsive materials and integrated techniques targeting the energy-efficient smart window application. *Advanced Functional Materials, 28*(22), 1800113.

[30] Makul, N., Fediuk, R., Amran, M., Zeyad, A. M., Murali, G., Vatin, N., . . . & Vasilev, Y. (2021). Use of recycled concrete aggregates in production of green cement-based concrete composites: A review. *Crystals, 11*(3), 232.

[31] Staniškis, J. K. (2010). Water saving in industry by cleaner production. In *Water Purification and Management* (pp. 1–33). Dordrecht: Springer Netherlands.

[32] Netravali, A. N. (Ed.). (2018). *Advanced green composites*. John Wiley & Sons.

[33] Ahmed, S. (2021). Advanced green materials: An overview. *Advanced green materials*, 1–13.

[34] Devarajan, Y. (2025). Nanomaterials-based wastewater treatment: Addressing challenges and advancing sustainable solutions. *BioNanoScience, 15*(1), 1–14.

[35] Wiederrecht, G. P., Bachelot, R., Xiong, H., Termentzidis, K., Nominé, A., Huang, J., . . . & Pupek, K. Z. (2023). Nanomaterials and sustainability.

[36] Palit, S., & Hussain, C. M. (2020). Green nanomaterials: a sustainable perspective. *Green nanomaterials: processing, properties, and applications*, 23–41.

[37] Suparna, M. G., & Rinsey Antony, V. A. (2016). Eco-friendly textiles. *International journal of science technology and management, 5*(11), 67–73.

[38] Novaković, M., Popović, D. M., Mladenović, N., Poparić, G. B., & Stanković, S. B. (2020). Development of comfortable and eco-friendly cellulose based textiles with improved sustainability. *Journal of Cleaner Production, 267*, 122154.

[39] Irimia-Vladu, M. (2014). "Green" electronics: biodegradable and biocompatible materials and devices for sustainable future. *Chemical Society Reviews, 43*(2), 588–610.

[40] Lovley, D. R. (2017). e-Biologics: fabrication of sustainable electronics with "green" biological materials. *MBio, 8*(3), 10–1128.

[41] Khan, Y., Thielens, A., Muin, S., Ting, J., Baumbauer, C., & Arias, A. C. (2020). A new frontier of printed electronics: flexible hybrid electronics. *Advanced Materials, 32*(15), 1905279.

[42] Bahadur, S., Sahu, U. K., Sahu, D., Sahu, G., & Roy, A. (2017). Review on natural gums and mucilage and their application as excipient. *Journal of applied pharmaceutical research, 5*(4), 13–21.

[43] van den Broek, L. A., Knoop, R. J., Kappen, F. H., & Boeriu, C. G. (2015). Chitosan films and blends for packaging material. *Carbohydrate polymers, 116*, 237–242.

[44] Gliessman, S. R. (2021). *Package price agroecology: The ecology of sustainable food systems*. CRC press.

[45] Bach, H., & Mauser, W. (2018). Sustainable agriculture and smart farming. In *Earth observation open science and innovation* (pp. 261–269). Cham: Springer International Publishing.

[46] Colwill, J. A., Wright, E. I., Rahimifard, S., & Clegg, A. J. (2012). Bio-plastics in the context of competing demands on agricultural land in 2050. *International Journal of Sustainable Engineering, 5*(1), 3–16.

4 Physicochemical Properties of Mucilage and Gums

4.1 Physicochemical Properties of Mucilage and Gums

Mucilage and gums are naturally occurring polysaccharides widely used in pharmaceuticals, food, and cosmetic industries due to their unique physicochemical properties. These hydrophilic substances can swell in the presence of water, forming viscous solutions or gels, making them effective as thickeners, stabilizers, and emulsifying agents [1]. Their molecular structure primarily consists of sugar units such as galactose, arabinose, mannose, and uronic acids, which contribute to their solubility and water-binding capacity. The viscosity of mucilage and gums varies, depending on the concentration, pH, temperature, and the presence of electrolytes, allowing customization for specific applications [1, 2]. They exhibit excellent film-forming abilities and bioadhesive characteristics, making them suitable for use in controlled drug delivery systems. Additionally, their colloidal nature enables them to stabilize suspensions and emulsions effectively [3]. These natural polymers are biodegradable and biocompatible, offering an environmentally friendly alternative to synthetic agents. Their rheological behavior, such as pseudoplasticity or shear-thinning properties, enhances their functionality in formulations where ease of flow under stress is desired. Moreover, mucilage and gums often possess antioxidant, antimicrobial, and anti-inflammatory properties, which further expand their utility in therapeutic formulations. Their multifunctional characteristics and natural origin make them valuable excipients and bioactive agents in modern formulations [2–4].

4.2 Chemical Composition of Mucilage

Mucilages are predominantly composed of polysaccharides (Figure 4.1), along with protein fractions, lipids, flavonoids, and minerals. Their composition can vary depending on the plant source, but several key components are commonly found in most mucilage [1, 4].

4.2.1 Polysaccharides [1]

Polysaccharides are the primary component of mucilage, forming long-chain carbohydrate polymers. The structure of these polysaccharides varies depending on the plant species, but they often consist of repeating units of hexoses, pentoses, and uronic acids.

https://doi.org/10.1515/9783111673509-004

Polysaccharides

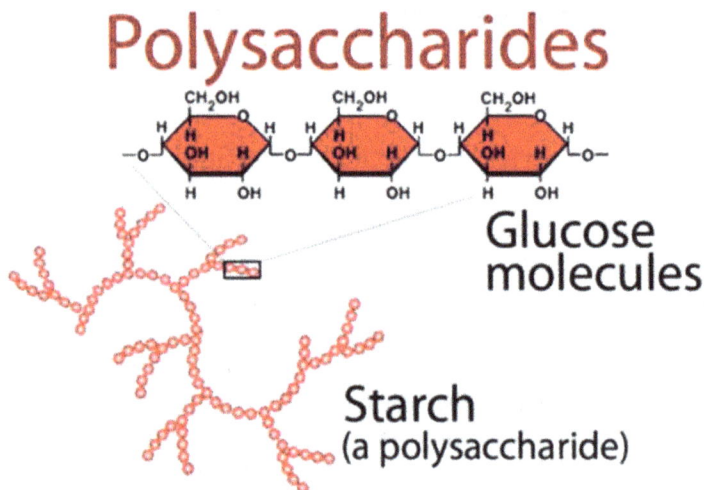

Glucose molecules

Starch
(a polysaccharide)

Figure 4.1: Chemical composition. Reproduced with permission from [4].

- **Galactose ($C_6H_{12}O_6$):** Galactose is often a major monosaccharide in mucilages. It is found in various galactose-rich mucilages such as guar gum and locust bean gum. Galactose forms a major structural component of mucilages and contributes to their water-binding and gelling properties.
- **Rhamnose ($C_6H_{12}O_5$):** Rhamnose is another significant sugar found in plant mucilages. It is often linked to other sugars such as galactose, arabinose, and xylose, contributing to the overall structure of the polysaccharide chains.
- **Arabinose ($C_5H_{10}O_5$):** Arabinose is a pentose sugar commonly present in mucilages, particularly in species like *Acacia*. It helps in forming highly branched polysaccharide structures.
- **Xylose ($C_5H_{10}O_5$):** Xylose, another pentose sugar, often appears in conjunction with arabinose in mucilage polysaccharides. It contributes to the structural integrity and water retention capacity of mucilage.
- **Uronic Acids:** Uronic acids, such as galacturonic acid ($C_6H_{10}O_7$) and glucuronic acid ($C_6H_{10}O_7$), are integral components of mucilage. These acids contribute to the hydrophilic nature of mucilage, enhancing its ability to absorb and retain water. Uronic acids are typically found as part of the polysaccharide structure, where they form ionic interactions with water molecules:
  - **Galacturonic acid** is primarily found in pectin-based mucilages.
  - **Glucuronic acid** plays a similar role in mucilages derived from species such as **flax** and **okra**.

4.2.2 Proteins and Amino Acids [2]

Mucilage contains small amounts of proteins, which may be in the form of glycoproteins. These proteins are crucial for stabilizing the mucilage structure and facilitating its interaction with other compounds. Common amino acids found in mucilage proteins include **glutamine, proline, glycine,** and **serine:**
–   **Glutamine** and **proline** are key in maintaining the stability of mucilage.
–   **Glycine** contributes to the flexibility of the polysaccharide chains.

4.2.3 Lipids [1, 2]

Though present in minor amounts, lipids contribute to the hydrophobicity of mucilage and aid in its emulsifying properties. Fatty acids like linoleic acid, oleic acid, and palmitic acid are often present.

4.2.4 Other Minor Components

Mucilage may contain a variety of bioactive compounds such as flavonoids, phenolic acids, and minerals like calcium (Ca^{2+}), potassium (K^+), magnesium (Mg^{2+}), and sodium (Na^+). These compounds influence the viscosity, gel formation, and overall behavior of the mucilage.

4.3 Chemical Composition of Gums

Gums are primarily composed of polysaccharides and can be classified into two categories: **exudate gums** (derived from plants) and **microbial gums** (produced by microorganisms). Their chemical composition generally includes monosaccharides, uronic acids, and, sometimes, proteins or lipids.

4.3.1 Polysaccharides [1, 5]

Gums are primarily polysaccharides that can either be linear or branched chains of simple sugars. The specific sugars involved vary depending on the type of gum.
–   **Galactose ($C_6H_{12}O_6$):** Galactose is commonly found in various gums such as guar gum, locust bean gum, and gum arabic. It is an essential sugar for the formation of the polysaccharide backbone.

- **Mannose ($C_6H_{12}O_6$):** Mannose is frequently seen in gums like guar gum, locust bean gum, and carob gum. Its presence contributes to the stability and texture of the gum.
- **Xylose ($C_5H_{10}O_5$):** Xylose is often present in gums such as guar gum and xanthan gum. Its interactions with water molecules help gums achieve their water-binding properties.
- **Rhamnose ($C_6H_{12}O_5$):** Rhamnose is found in several gums, including gum arabic and xanthan gum. It plays a role in the gum's ability to form gel-like structures.
- **Arabinose ($C_5H_{10}O_5$):** Arabinose is another common sugar in gums such as **gum arabic**. It contributes to the gum's gel-forming ability.
- **Uronic Acids:** Uronic acids like **galacturonic acid** and **glucuronic acid** play an important role in the water retention and gelation characteristics of gums.

4.3.2 Proteins and Glycoproteins [6]

Some gums, like **gum arabic**, contain a significant amount of protein (10–15% by weight). These proteins are primarily **glycoproteins**, with carbohydrate chains attached to the protein backbone. The presence of proteins in gums contributes to their stability, emulsification, and water retention properties:
- Guar gum, for example, contains 1–3% protein, which adds to its ability to thicken liquids.

4.3.3 Minor Components

Like mucilage, gums also contain minor compounds that influence their properties. These include lipids, flavonoids, phenolic acids, and other bioactive substances:
- Xanthan gum, for example, contains small amounts of acetate and pyruvate groups attached to the polysaccharide backbone, which impact its rheological properties (flow and deformation characteristics) [7].

4.4 Viscosity and Gelation Mechanism

The viscosity and gelation properties of mucilages and gums are closely related to their chemical composition, particularly the structure of the polysaccharides. These substances have the unique ability to form gels or viscous solutions when dissolved in water.

4.4.1 Mucilage Viscosity [8]

Mucilages are highly viscous due to the high content of polysaccharides and uronic acids. These components form hydrogen bonds with water molecules, leading to a thickening effect. The presence of uronic acids (such as galacturonic acid and glucuronic acid) promotes cross-linking, which further enhances water retention and gel formation. This is critical for the roles mucilage plays in plants, such as seed protection and drought resistance.

4.5.2 Gum Viscosity [9]

Gums exhibit similar viscosifying properties. For example, guar gum and xanthan gum are known for their excellent water-binding and thickening abilities. These gums can form gels when dissolved in water, owing to the interactions between the sugar residues and water molecules. The gelation process is often influenced by factors such as temperature, pH, and the presence of ions like calcium (Ca^{2+}). **Xanthan gum**, for instance, has side chains that include mannose and glucuronic acid. These side chains interact with water to form a highly hydrated structure, leading to its well-known high viscosity. When calcium ions are present, they can form cross-links with glucuronic acid, aiding in gel formation.

4.5 Rheological Properties of Mucilage [8]

Mucilage is a complex polysaccharide network that could retain large amounts of water. The rheological behavior of mucilage is primarily determined by its **viscosity**, **elasticity**, and **gel formation properties**, which depend on the type of mucilage, its concentration, and the presence of other components, such as proteins and lipids.

4.5.1 Viscosity

Viscosity is a measure of a fluid's resistance to flow. Mucilage exhibits non-Newtonian behavior, meaning its viscosity changes under different shear rates, as shown in Table 4.1. This behavior is typically shear-thinning, where the viscosity decreases as shear stress increases. This is often observed in mucilage from plants such as okra, flaxseed, and cactus. The viscosity of mucilage can be quantified using a viscometer at different shear rates. A typical shear-thinning curve shows that at low shear rates, mucilage is more viscous, while at high shear rates, it becomes more fluid.

In this case, as the shear rate increases, the viscosity decreases, indicating a shear-thinning behavior, typical of mucilage.

Table 4.1: Example data for okra mucilage.

Shear rate (s^{-1})	Viscosity (Pa.s)
1	0.055
10	0.025
100	0.015
1,000	0.010

4.5.2 Elasticity and Gel Formation [10]

Mucilage is known for its ability to form gels. These gels are primarily hydrophilic due to the high water content in the mucilage. When mucilage interacts with water, it swells and forms a network structure that gives it elasticity. The elastic modulus (G'), which is the measure of the material's elasticity, can be evaluated through oscillatory shear measurements. For example, flaxseed mucilage has been shown to exhibit gelation at concentrations above 2% (w/v). When subjected to an oscillatory shear test, flaxseed mucilage demonstrates a gel-like response at higher concentrations, with $G' > G''$, where G' is the storage modulus (elastic component), and G'' is the loss modulus (viscous component). At higher concentrations of mucilage, such as 3%, the elastic modulus (G') increases significantly, indicating that the material is more solid and elastic. These properties are key for mucilages used as thickening agents or gelatinous substances in food.

4.6 Rheological Behavior of Gums [11]

Gums are another class of polysaccharides that exhibit unique rheological behaviors. Unlike mucilage, gums are often used in food and industrial applications for their ability to increase viscosity, stabilize emulsions, and form gels. Gums can be derived from both plant exudates and microbial sources, each displaying specific rheological characteristics based on their chemical structure and molecular interactions.

4.6.1 Viscosity of Gums [12]

Gums are commonly used to increase the viscosity of aqueous solutions. Their behavior in water is largely influenced by the concentration, type of gum, and shear rate. Like mucilage, most gums exhibit shear-thinning behavior. Guar gum, shown in Table 4.2, is widely used in the food industry because of its high water-binding capacity.

Table 4.2: Example data for guar gum.

Shear rate (s^{-1})	Viscosity (Pa.s)
1	0.10
10	0.05
100	0.02
1,000	0.01

The viscosity of guar gum decreases as shear rate increases, following the shear-thinning behavior. The effect of concentration on viscosity is also significant; increasing the concentration of guar gum leads to an increase in viscosity.

4.6.2 Gelation and Elasticity [10, 11]

Gums such as xanthan gum and gellan gum have a remarkable ability to form gels, especially in the presence of divalent cations such as calcium. The gelation process involves the formation of a network structure that traps water and creates a stable gel. The elastic modulus (G') for gums can also be measured via oscillatory rheology. At higher concentrations of xanthan gum, the elastic modulus increases, indicating that the gum forms a stronger gel with higher elasticity.

4.6.3 Effect of Temperature and pH on Gums [13]

Gums are highly sensitive to temperature and pH. The rheological behavior of gums, especially their viscosity and gel strength, can change significantly under varying temperatures or pH conditions:

- **Xanthan Gum**: Its viscosity remains stable over a wide range of temperatures and pH levels. However, at temperatures above **70 °C**, its viscosity may decrease, and **gelation** can be disrupted at lower pH values [7].
- **Guar Gum**: **Guar gum**'s viscosity decreases with increasing temperature, which is characteristic of many **polysaccharide-based** gums [7].

Guar gum loses viscosity as temperature increases, a typical behavior for many polysaccharides in aqueous solutions.

4.7 Structural Characterization [14–17]

4.7.1 Fourier-Transform Infrared (FTIR) Spectroscopy

Fourier-transform infrared (FTIR) spectroscopy is one of the most widely used techniques to identify functional groups and chemical bonds in polysaccharides, including mucilage and gums. FTIR measures the absorption of infrared light by the material as a function of frequency, providing a spectral fingerprint that can be linked to specific molecular vibrations. In the case of mucilage and gums, FTIR can identify key functional groups like hydroxyl (OH), carbonyl (C = O), ether (C–O–C), and amino (NH_2) groups that are characteristic of polysaccharides shown in Table 4.3.

Table 4.3: Example FTIR spectrum for guar gum.

Wavelength (cm^{-1})	Absorption band (cm^{-1})	Assignment
3,400–3,200	3,400–3,200	O–H stretching (hydroxyl)
2,950–2,800	2,920–2,850	C–H stretching (methyl)
1,650	1,640	C = O stretching (carbonyl)
1,150–1,000	1,060	C–O–C stretching (ether bond)

For guar gum, the O–H stretch of around 3,400 cm^{-1} indicates the presence of hydroxyl groups, which are highly hydrophilic and contribute to the water-binding ability of the gum. The C–H stretch in the region between 2,950–2,800 cm^{-1} is due to the methyl groups in the polysaccharide structure, which are part of the side chains in guar gum. FTIR also reveals the ether linkages (C–O–C) at 1,060 cm^{-1}, typical of the polysaccharide backbone structure, and the carbonyl stretch at 1,640 cm^{-1}, which indicates the presence of uronic acid residues, common in gums such as guar and xanthan.

4.7.2 Nuclear Magnetic Resonance (NMR) Spectroscopy

Nuclear magnetic resonance (NMR) spectroscopy provides a more detailed understanding of the molecular structure of mucilage and gums, including the specific monosaccharides and their arrangement in the polymer chain. Through 1H-NMR and 13C-NMR, the chemical shifts of hydrogen and carbon atoms reveal the connectivity between sugar units, side chains, and functional groups (Table 4.4).

For xanthan gum, 1H-NMR analysis typically shows peaks corresponding to the β-glucose backbone, with a signal in the 5.2–4.9 ppm range for the anomeric proton. The 3.8–3.6 ppm region corresponds to protons involved in glycosidic linkages, while

Table 4.4: Example NMR data for xanthan gum.

Chemical shift (ppm)	Assignment
5.2–4.9	H–C–H (anomeric proton, β-glucose)
3.8–3.6	H–C–H (glycosidic bond protons)
2.1–1.8	C–H (side chain acetyl groups)

the 2.1–1.8 ppm region reflects the acetyl groups on the side chains, which are important for the gel formation and viscosity of xanthan gum.

Moreover, 13C-NMR can be employed to discern the specific carbon atom environments, allowing researchers to differentiate between types of sugars (glucose, mannose, rhamnose) and their linkages within the polymer structure.

4.7.3 Raman Spectroscopy

Raman spectroscopy is another vibrational technique used to investigate the molecular structure of mucilage and gums. Unlike FTIR, which measures absorption, Raman spectroscopy measures the scattering of light, providing complementary information on the vibrational modes of the chemical bonds.

Raman spectroscopy is particularly useful for studying the **stereochemistry** of sugar units and the **configuration** of glycosidic linkages in mucilage and gums.

4.7.4 UV-Vis Spectroscopy

UV-Vis spectroscopy is typically used to examine the aromatic components and the conjugated systems in polysaccharides, although its application in mucilages and gums is less common. However, certain gum and mucilage samples, particularly those derived from plants containing phenolic compounds, may exhibit UV absorption peaks that provide insights into their secondary metabolites.

4.7.5 Scanning Electron Microscopy (SEM)

Scanning electron microscopy (SEM) is a powerful tool for morphological characterization of mucilage and gums. SEM can provide high-resolution images of the surface structures, size, and distribution of polysaccharide particles and gel networks. SEM is particularly useful in understanding the microstructure of dried gum exudates and mucilage gels.

For example, **guar gum** and **xanthan gum** exhibit **rough** and **porous** surfaces, which are important for their water absorption properties:

- **Xanthan Gum**: The surface of xanthan gum under SEM shows a **gel-like structure** with interconnected **fibrous network** morphology, indicating its high water **retention** capacity.
- **Guar Gum**: Guar gum presents a **granular structure** with **irregular** shapes that swell when hydrated, forming a **gel-like matrix**.

4.7.6 Transmission Electron Microscopy (TEM)

Transmission electron microscopy (TEM) is used for higher-resolution imaging, allowing researchers to examine the internal structure of polysaccharide gels and the ultrastructure of polymer chains. TEM is particularly useful for studying the nanostructure of mucilage films and their interactions with water molecules:

- For example, xanthan gum forms a network of fibrils that can be visualized under TEM, which are essential for its viscosity and gelation properties.

4.7.7 Atomic Force Microscopy (AFM)

Atomic force microscopy (AFM) is a scanning probe microscopy technique that provides topographical images with nanoscale resolution. AFM is valuable for studying the surface roughness, elasticity, and nanoscale interactions of mucilages and gums. For instance, AFM can reveal the structural heterogeneity of gum arabic or guar gum in its gel form, offering insights into how the gum interacts with its environment on the nanometer scale.

4.7.8 X-Ray Diffraction (XRD)

X-ray diffraction (XRD) is employed to examine the crystalline structure of gums and mucilage. Many gums, such as guar gum, exhibit amorphous or semi-crystalline structures, depending on the processing conditions. XRD analysis, mentioned in Table 4.5, helps determine the degree of crystallinity, which influences the solubility, viscosity, and gelation properties of these materials.

The peak intensities at specific 2θ angles indicate the crystallinity of the sample. Guar gum, for instance, shows a low to medium degree of crystallinity, which means it is largely amorphous but may contain some crystalline domains that influence its viscoelastic properties.

Table 4.5: XRD patterns of guar gum.

Angle (2θ)	Peak intensity (a.u.)	Crystallinity
15°	1.5	Low
20°	2.0	Medium
30°	3.5	High

4.7.9 Gel Permeation Chromatography (GPC)

Gel permeation chromatography (GPC), also known as size-exclusion chromatography, is used to determine the molecular weight distribution and polydispersity of mucilages and gums. This technique separates molecules based on their size, allowing researchers to assess the polymeric nature of the gums and their ability to form gel structures.

For example, xanthan gum typically exhibits a high molecular weight (ranging from 10^5 to 10^7 g/mol), which contributes to its viscosity and gelation properties. GPC analysis can confirm the presence of high-molecular-weight fractions, which are essential for the functional behavior of these materials.

References

[1]   Tosif, M. M., Najda, A., Bains, A., Kaushik, R., Dhull, S. B., Chawla, P., & Walasek-Janusz, M. (2021). A comprehensive review on plant-derived mucilage: characterization, functional properties, applications, and its utilization for nanocarrier fabrication. *Polymers*, *13*(7), 1066.

[2]   Agarwal, S., & Upadhyay, P. Synthesis, Optimization and Characterization Of Abelmoschus Esculentus Mucilage As A Potential New Age Nano-Therapeutic Intervention.

[3]   Bhat, U. R., & Tharanathan, R. N. (1987). Functional properties of okra (Hibiscus esculentus) mucilage. *Starch-Stärke*, *39*(5), 165–167.

[4]   Aly, A. A., & El-Bisi, M. K. (2018). Grafting of polysaccharides: Recent advances. Biopolymer grafting, 469–519.

[5]   Andrade, L. A., da Silva, A. C., & Pereira, J. (2024). Chemical composition of taro mucilage from different extraction techniques found in literature. *Food Chemistry Advances*, *4*, 100648.

[6]   Gleeson, P. A., & Clarke, A. E. (1979). Structural studies on the major component of Gladiolus style mucilage, an arabinogalactan-protein. *Biochemical Journal*, *181*(3), 607–621.

[7]   Sworn, G. (2021). Xanthan gum. In *Handbook of hydrocolloids* (pp. 833–853). Woodhead Publishing.

[8]   Medina-Torres, L., Brito-De La Fuente, E., Torrestiana-Sanchez, B., & Katthain, R. (2000). Rheological properties of the mucilage gum (Opuntia ficus indica). *Food hydrocolloids*, *14*(5), 417–424.

[9]   Doublier, J. L., Garnier, C., & Cuvelier, G. (2017). Gums and hydrocolloids: functional aspects. In *Carbohydrates in food* (pp. 307–354). CRC Press.

[10] Medina-Torres, L., Brito-De La Fuente, E., Torrestiana-Sanchez, B., & Alonso, S. (2003). Mechanical properties of gels formed by mixtures of mucilage gum (Opuntia ficus indica) and carrageenans. *Carbohydrate polymers*, *52*(2), 143–150.

[11]   Chenlo, F., Moreira, R., & Silva, C. (2010). Rheological behaviour of aqueous systems of tragacanth and guar gums with storage time. *Journal of Food Engineering, 96*(1), 107–113.

[12]   Casas, J. A., Mohedano, A. F., & García-Ochoa, F. (2000). Viscosity of guar gum and xanthan/guar gum mixture solutions. *Journal of the Science of Food and Agriculture, 80*(12), 1722–1727.

[13]   Brunchi, C. E., Bercea, M., Morariu, S., & Dascalu, M. (2016). Some properties of xanthan gum in aqueous solutions: effect of temperature and pH. *Journal of Polymer Research, 23*(7), 123.

[14]   Singh, S., & Bothara, S. B. (2014). Physico-chemical and structural characterization of mucilage isolated from seeds of Diospyros melonoxylon Roxb. *Brazilian Journal of Pharmaceutical Sciences, 50*(4), 713–725.

[15]   Singh, R., & Barreca, D. (2020). Analysis of gums and mucilages. In *Recent Advances in Natural Products Analysis* (pp. 663–676). Elsevier.

[16]   Singh, S., & Bothara, S. (2014). Morphological, physico-chemical and structural characterization of mucilage isolated from the seeds of Buchanania lanzan Spreng. *International Journal of Health & Allied Sciences, 3*(1), 33–33.

[17]   Granzotto, C. (2014). Methodological developments based on mass spectrometry for the analysis of glycoproteins and polysaccharides of plant gums: an application to cultural heritage samples.

5 Functional Properties and Applications

5.1 Introduction

Natural polymers and biopolymers exhibit vital functional properties such as gelling, emulsifying, and thickening that are extensively harnessed in food, pharmaceutical, and cosmetic industries. Their gelling ability allows the formation of three-dimensional networks capable of trapping water, which is crucial in producing structured gels and controlled-release drug systems [1]. Emulsifying properties facilitate the stabilization of oil-water mixtures, vital in food emulsions, topical formulations, and suspensions. These polymers also act as excellent thickening agents, modifying the consistency and texture of products such as sauces, creams, and oral suspensions [2]. Solubility plays a critical role in determining the dispersion and application of these polymers in aqueous systems, influencing product clarity and stability. Viscosity, which refers to the resistance to flow, directly impacts processing and sensory attributes in food and pharmaceutical formulations. The hydration behavior, or the ability of the polymer to absorb and retain water, is fundamental in maintaining moisture content, improving shelf life, and controlling release mechanisms in drug delivery. These functional properties collectively enable natural polymers to enhance product stability, bioavailability, and user acceptability, underscoring their significance as multifunctional excipients across various formulations and industrial applications [3, 4].

5.2 Gelling Properties

Gelling agents are substances that are used to form a gel-like structure when combined with water or other solvents. Gels are semisolid systems that exhibit unique viscoelastic behavior, which means they show both elastic (solid-like) and viscous (liquid-like) characteristics. Gels are formed when polymeric molecules interact with water molecules, forming a network structure that traps water and other ingredients within it. These networks can be made more or less rigid by adjusting factors such as the concentration of gelling agent, temperature, and pH [5].

5.2.1 Common Gelling Agents

- **Gelatin:** Gelatin is one of the most well-known gelling agents, derived from animal collagen. It is widely used in confectionery products, jellies, gummies, and marshmallows. The gelling strength of gelatin is quantified in **bloom strength**, a measure of its gel strength, with higher bloom values indicating stronger gels.

https://doi.org/10.1515/9783111673509-005

Typical bloom strength ranges from **100 to 250**. For example, a gelatin with bloom strength of **200** would be stronger than one with a bloom of **150** [6].

– **Agar-Agar:** This gelling agent is derived from red algae and is often used as a vegan alternative to gelatin. Agar has a higher melting point than gelatin, typically forming gels at temperatures as high as 85 °C (185 °F), compared to gelatin, which gels at 30–35 °C (86–95 °F). The gelling strength of agar is usually quantified in terms of gel strength in g/cm^2, where values typically range from 500–1,000 g/cm^2 for different agar formulations [7].

– **Pectin:** Found in fruits, pectin is frequently used in making jams, jellies, and marmalades. Pectin works by forming a gel network when it is combined with sugars and acids. The degree of gelling can be influenced by the sugar-to-acid ratio. The gelling strength of pectin can vary but is typically defined by the gelation temperature, which ranges between 40 and 70 °C (104–158 °F) depending on the type of pectin used [8].

The gelling process is influenced by several factors [7, 8]:

– **Concentration:** A higher concentration of gelling agent leads to a firmer gel. For example, the concentration of gelatin needed to form a firm gel is typically in the range of **5–10%** by weight.

– **Temperature:** Gels made with gelatin or agar are temperature-sensitive. Gelatin gels at lower temperatures (below 35 °C), while agar requires a higher temperature to form a gel and can withstand higher temperatures without melting, making it more useful in hot applications.

– **pH and Ionic Strength:** The pH of the solution can influence the gelling ability of some gelling agents, such as pectin. An acidic pH (around 3.0–3.5) is optimal for pectin gelation. Similarly, the presence of salts, like calcium ions in pectin, can promote gelation, which is a key factor in making jams and jellies.

Applications of gelling agents are widespread across various industries, including food (for desserts, candies, and sauces), pharmaceuticals (for controlled-release drug delivery), and biotechnology (for cell culture matrices). The numerical values associated with gel strength and gelling temperature guide the selection of the appropriate agent for a specific application.

5.3 Emulsifying Properties

Emulsification is the process of dispersing one liquid into another immiscible liquid, such as oil in water, to form a stable emulsion. Emulsifiers are agents that reduce the surface tension between two immiscible liquids and stabilize the dispersion of droplets. They achieve this by having both hydrophilic (water-loving) and hydrophobic (oil-loving) regions in their molecular structure, allowing them to bridge the gap between water and oil phases [9].

5.3.1 Common Emulsifiers

- **Lecithin:** This is one of the most widely used natural emulsifiers, typically extracted from soybeans or egg yolk. Lecithin contains phospholipids, which have both hydrophilic and hydrophobic portions, making it effective in stabilizing oil-in-water emulsions. Lecithin is commonly used in products like mayonnaise, salad dressings, and margarine. Its HLB (hydrophilic-lipophilic balance) value typically ranges from 8 to 12, depending on the source and composition. An HLB value between 8 and 10 is optimal for forming stable emulsions in food systems [10].
- **Monoglycerides and Diglycerides:** These are derived from fats and oils and are widely used in the baking industry to improve the texture and shelf life of bread and cakes. Monoglycerides are particularly effective in stabilizing emulsions in ice cream and dairy products. The concentration of monoglycerides used typically ranges from **0.5% to 2%** in food products, depending on the desired emulsification strength [11].
- **Polysorbates (Tween 20, Tween 80):** Polysorbates are synthetic emulsifiers used in a wide range of food and pharmaceutical applications. Tween 80, for instance, is used to stabilize emulsions in cosmetic and pharmaceutical formulations. The concentration of polysorbates needed for effective emulsification is typically around 0.1–1% [12].

The efficiency of emulsifiers is often measured by the droplet size of the dispersed phase. Smaller droplets (typically 1–10 microns) contribute to more stable emulsions and are often preferred in high-quality food products like mayonnaise and salad dressings. Additionally, the emulsion stability can be quantified using parameters such as creaming rate and phase separation. For example, a well-emulsified mayonnaise should show minimal phase separation even after prolonged storage.

Applications of Emulsifiers: The food industry extensively uses emulsifiers to improve the texture, consistency, and shelf life of products. In pharmaceuticals, emulsifiers help in creating stable formulations for oral and topical drugs. Emulsifiers are also critical in cosmetic products, where they enable the smooth mixing of water and oil-based ingredients.

5.4 Thickening Properties

Thickening agents are substances that increase the viscosity of a liquid without substantially altering its other properties. The primary function of thickeners is to modify the texture of food products, enhancing mouthfeel and improving stability. Thickeners typically act by increasing the number of interactions between molecules within the liquid, thereby making it more resistant to flow [13].

5.4.1 Common Thickening Agents

- **Starches:** Starch is one of the most common and widely used thickening agents in food products. It can be derived from corn, potato, wheat, and tapioca. Starch granules absorb water and swell when heated, leading to an increase in viscosity. The thickening power of starches is often measured in terms of gelatinization temperature, which typically ranges from 60 to 80 °C (140–176 °F) for most starches. The concentration needed to achieve a desired viscosity is typically between 1–10% depending on the specific application [14].
- **Xanthan Gum:** This is a polysaccharide produced by fermentation of carbohydrates. It is widely used as a thickening agent in gluten-free baking, salad dressings, and sauces. Xanthan gum has the ability to increase the viscosity of liquids even at low concentrations, typically ranging from **0.1% to 0.5%**. Xanthan gum's ability to form stable gels makes it ideal for use in products that require consistency at varying temperatures and conditions [15].
- **Guar Gum:** Another popular natural thickener, guar gum is derived from the seeds of the guar plant. It is primarily used in dairy, bakery, and processed food products to improve texture. The viscosity of guar gum solutions increases significantly even at low concentrations, typically ranging from **0.1% to 1.0%** [16].
- **Carrageenan:** This seaweed-derived polysaccharide is commonly used in dairy products like ice cream and puddings to provide a smooth, creamy texture. The thickening properties of carrageenan depend on the type used (kappa, iota, or lambda), with kappa-carrageenan typically forming firmer gels and thicker solutions. The concentration required for thickening is generally around 0.5–2.0% [17].

The thickening ability of these agents is often determined by viscosity measurements using a viscometer. Viscosity is typically expressed in centipoise (cP) or Pascal-seconds (Pa·s). For instance, a typical starch-based pudding may have a viscosity of around 500–1,000 cP at room temperature, whereas a thickened sauce might have a viscosity of 2000–4,000 cP.

Thickeners can also impact the suspension stability of products. For example, thickeners are used to prevent the separation of solid particles in sauces and soups, maintaining uniformity over time. The ability of thickeners to modify the yield stress (the minimum stress required to initiate flow) is important in food processing, particularly for products like dressings, sauces, and gravies.

5.4.2 Applications of Thickeners

Thickeners are essential in creating desirable textures in a wide range of products, from soups and sauces to processed meats and dairy products. In the pharmaceutical industry, thickeners are used in suspensions and oral formulations to ensure proper drug delivery. Thickeners are also employed in personal care products such as lotions, shampoos, and creams, where they contribute to the product's texture and stability.

Gelling, emulsifying, and thickening agents play essential roles in the formulation and stability of food and nonfood products. Each of these functional properties has been optimized through years of research, and their applications are diverse and widespread. The numerical values associated with these agents such as gel strength, emulsion droplet size, and viscosity serve as important indicators of their performance and guide their use in specific applications [5].

5.5 Solubility, Viscosity, and Hydration Behavior

Solubility is the ability of a substance (solute) to dissolve in a solvent (usually water) to form a homogeneous solution. This property is crucial in a wide range of applications, from pharmaceutical drug formulations to food and beverage processing. The solubility of a substance depends on several factors, including temperature, pressure, the nature of the solute and solvent, and the presence of other components in the system (such as salts or pH modifiers) [18].

The solubility product constant (K_{sp}) is a measure of the solubility of a compound in solution and is particularly relevant for sparingly soluble substances. The K_{sp} value indicates how well a compound dissolves in water at a specific temperature. For instance, the solubility product for calcium sulfate ($CaSO_4$) at 25 °C is approximately 4.93×10^{-5} mol^2/L^2, which means that the compound dissolves only slightly in water at room temperature [19].

5.5.1 Factors Influencing Solubility [18]

– **Temperature:** Solubility generally increases with temperature for most solid solutes. For example, the solubility of sucrose in water increases from approximately 200 g/100 mL at 0 °C to 1,000 g/100 mL at 100 °C.

- **Pressure:** Solubility is more relevant for gases, as the solubility of gases in liquids increases with an increase in pressure. This relationship is described by Henry's law. For example, at pressure of 1 atm, the solubility of oxygen in water is about 0.003 mg/L at **0 °C**, but this increases when the pressure is higher.
- **Nature of the Solute and Solvent:** Polar substances generally dissolve well in polar solvents, like water, whereas nonpolar substances tend to dissolve in nonpolar solvents. For example, **salt** (sodium chloride) dissolves in water due to the polarity of both the salt and the water molecules, while **oil** does not dissolve in water due to their nonpolar natures.
- **Presence of Other Components:** Some solutes can affect the solubility of other substances. For instance, the solubility of calcium carbonate (**$CaCO_3$**) is decreased in water with high concentrations of dissolved **CO_2** due to the formation of calcium bicarbonate.

5.5.2 Applications of Solubility

- **Pharmaceuticals:** In drug design, solubility is critical for ensuring proper absorption of active pharmaceutical ingredients (APIs). Many drugs are poorly soluble in water, limiting their bioavailability. For example, drugs like ibuprofen have a low solubility in water, which has led to the development of solubility-enhancing formulations such as salt forms, liposomal encapsulation, or nanosizing techniques [20].
- **Food and Beverages:** Solubility affects the quality of many food products, such as beverages, sauces, and seasonings. For example, sugar and salt are highly soluble in water, which is why they dissolve quickly in drinks or soups. On the other hand, substances like chocolate powder or coffee grounds exhibit limited solubility, requiring stirring or agitation to aid dissolution [21].
- **Cosmetics and Personal Care:** Solubility also plays a crucial role in the formulation of skincare products. Water-soluble ingredients like hyaluronic acid can hydrate the skin, while oil-soluble ingredients like vitamin E are used for their antioxidant properties [22].

5.6 Viscosity

Viscosity refers to the resistance of a fluid to flow. It is a measure of the internal friction within a fluid as molecules or particles interact with each other. Higher viscosity means the fluid flows more slowly (thicker fluid), while lower viscosity corresponds to a fluid that flows more easily (thinner fluid). Viscosity is a crucial property in industries like food processing, paints, pharmaceuticals, and lubrication. The viscosity of a liquid can be measured using instruments like a **viscometer** or a **rheometer**.

The unit of viscosity is typically Pascal-seconds (Pa·s) or centipoise (cP), where 1 cP equals 0.001 Pa·s [23].

5.6.1 Factors Influencing Viscosity [5, 23]

- **Temperature:** Viscosity decreases as temperature increases for most liquids. This occurs because higher temperatures cause molecules to move faster, reducing the internal friction between them. For example, the viscosity of water at **20 °C** is **1.002 cP**, but at **100 °C**, it decreases to about **0.282 cP**.
- **Concentration:** The concentration of solutes can significantly impact viscosity. In solutions, increasing the concentration of dissolved substances like starch or xanthan gum typically leads to an increase in viscosity. For instance, a **1%** solution of **xanthan gum** in water has a viscosity of about **10 cP**, while a **3%** solution may have a viscosity of around **100 cP**.
- **Shear Rate:** Many fluids, particularly those that exhibit non-Newtonian behavior, show a change in viscosity depending on the rate at which they are sheared. **Shear-thinning fluids** (e.g., ketchup, sauces) decrease in viscosity as they are stirred or pumped, while **shear-thickening fluids** (e.g., cornstarch slurry) increase in viscosity under shear stress.
- **Molecular Structure:** The viscosity of a fluid is also influenced by the size and shape of its molecules. Long-chain polymers or molecules with high molecular weight contribute to higher viscosity. For example, **polymers** such as **polyethylene glycol** exhibit high viscosity due to their long molecular chains.

5.6.2 Applications of Viscosity

- **Food Industry:** Viscosity plays a crucial role in determining the texture and mouthfeel of food products. For example, the viscosity of sauces, soups, and dressings directly influences consumer perceptions of quality. In beverages like smoothies or milkshakes, controlled viscosity gives the product a desirable thickness, which can be adjusted by adding thickeners such as **guar gum** or **xanthan gum** [24].
- **Pharmaceuticals:** In the pharmaceutical industry, the viscosity of syrups, suspensions, and creams must be controlled to ensure proper drug delivery. For example, the viscosity of **liquid medicines** affects their ease of swallowing, as higher viscosity may be more difficult to consume. Similarly, topical creams must have the right viscosity for ease of application and absorption [25].
- **Lubricants and Coatings:** The viscosity of oils and lubricants affects their performance in machinery. For instance, motor oils with low viscosity (e.g., SAE 5W-30) are designed for colder temperatures and fast-flowing, low-resistance lubrication,

while oils with higher viscosity (e.g., SAE 20W-50) are used in warmer conditions for thicker lubrication [26].

5.7 Hydration Behavior

Hydration behavior refers to the ability of a substance to absorb water and swell when in contact with it. Hydration is crucial in many applications, including the preparation of food products, drug formulations, and the development of materials with specific structural or functional properties. The hydration behavior of a substance is influenced by its molecular structure, the presence of functional groups, and the interactions with water molecules [27].

Hydration is an essential process in the formation of gels, pastes, and thickened solutions. The rate of hydration, the final water content, and the resulting consistency are key characteristics for many applications.

5.7.1 Key Parameters of Hydration Behavior [27]

– **Hydration Capacity:** This refers to the maximum amount of water a substance can absorb. It is usually expressed as the gram of water per gram of dry substance (g/g). For example, guar gum has a hydration capacity of about 7–8 g of water per gram of gum powder, making it highly effective as a thickener.
– **Water Absorption Rate:** The rate at which a substance hydrates can vary significantly. For example, gelatin hydrates relatively quickly in cold water and forms a gel at relatively low temperatures (30–35 °C), while agar requires boiling water to fully hydrate and form a gel.
– **Swelling Power:** This is a measure of the volume increase a substance undergoes upon hydration, typically expressed as the increase in volume (mL or cm^3) per gram of the substance. For example, starch can swell significantly when hydrated, with swelling powers ranging from 5 to 10 mL/g, depending on the type of starch.
– **Gelation:** Hydration also influences the ability of certain substances to form gels. For instance, agar forms a gel after hydration and cooling, whereas pectin requires both hydration and the presence of sugar and acid to form a stable gel.

5.7.2 Applications of Hydration Behavior [27]

Food Industry: Hydration behavior is vital in the preparation of products such as bread, cakes, and pastas. The ability of flour to absorb water influences dough consistency, while the hydration of gelatin or pectin plays a role in producing gummy candies, jellies, and other gel-based foods.

Pharmaceuticals: Hydration is critical in the formulation of controlled-release tablets and capsules. Polymers such as hydroxypropyl methylcellulose (HPMC) absorb water and swell to form a gel matrix that controls the release of drugs. Hydration also impacts the stability and efficacy of suspensions and emulsions.

Agriculture: Hydration is a key property in the formulation of hydrocolloid-based soil conditioners and fertilizers. Hydrogels such as polyacrylamide absorb large quantities of water and help retain moisture in soil, improving irrigation efficiency.

5.8 Biological Activities [28]

Biological activities of various substances, particularly their antioxidant and antimicrobial properties, play a pivotal role in enhancing human health and promoting the efficacy of therapeutic and preventive measures across several industries.

5.8.1 Antioxidant Activity: Mechanism and Importance

Antioxidants are molecules that prevent or slow down the oxidation process by neutralizing unstable molecules of free radicals, which can damage cellular structures. Free radicals, which include reactive oxygen species (ROS) like superoxide anion (O_2^-), hydroxyl radicals (OH•), and hydrogen peroxide (H_2O_2), are naturally produced during cellular metabolism but can also result from environmental stressors such as UV radiation, pollution, and smoking [29].

5.8.1.1 Methods of Measuring Antioxidant Activity
Several laboratory techniques are used to measure the antioxidant activity of substances:
1. **DPPH Assay**: This is a popular method for assessing the radical-scavenging activity of antioxidants. The DPPH radical (2,2-diphenyl-1-picrylhydrazyl) is a stable free radical that, when reduced by an antioxidant, causes a decrease in absorbance at 517 nm. The IC_{50} value, which represents the concentration of antioxidant required to reduce DPPH by 50%, is commonly used to quantify antioxidant activity. Lower IC_{50} values indicate higher antioxidant potency. For example, vitamin C has an IC_{50} of approximately 10 µg/mL, while green tea polyphenols may have an IC_{50} around 50 µg/mL, indicating that vitamin C is a stronger antioxidant [30].
2. **ABTS Assay**: The ABTS (2,2'-azinobis-(3-ethylbenzothiazoline-6-sulfonic acid)) assay measures the ability of antioxidants to neutralize the ABTS radical cation. The results are expressed as the Trolox equivalent antioxidant capacity (TEAC), where Trolox, a water-soluble vitamin E analogue, is used as a standard. The TEAC value for a substance indicates its ability to scavenge ABTS radicals relative

to Trolox. For instance, grapeseed extract might have a TEAC value of around 2 mmol Trolox/g, whereas vitamin E might show a value of 1 mmol Trolox/g [31].

3. **FRAP Assay**: The ferric reducing antioxidant power assay is based on the reduction of Fe^{3+} to Fe^{2+} by antioxidants. The FRAP value is calculated as the concentration of antioxidant equivalent to the reduction of Fe^{3+} in micromoles per gram (μmol/g). For example, blackberries have an FRAP value of about 6–7 mmol Fe^{2+}/100 g, while green tea can exhibit values upwards of 50 mmol Fe^{2+}/100 g, showing a higher antioxidant power [32].

4. **ORAC Assay**: The oxygen radical absorbance capacity assay measures the ability of a sample to protect against peroxyl radicals. The result is expressed in μmol Trolox equivalents per gram or liter. Spinach and blueberries are known for high ORAC values, which can range from 3,000 to 8,000 μmol TE/g [33].

5.8.1.2 Factors Influencing Antioxidant Activity [31, 33]

1. **Concentration**: The concentration of antioxidants plays a crucial role in their effectiveness. For example, at low concentrations, antioxidants may exhibit **prooxidant** behavior, particularly metal-chelating antioxidants like **iron** or **copper**.

2. **pH**: The antioxidant activity of many compounds is pH-dependent. For example, **flavonoids** have different antioxidant capacities at acidic and neutral pH levels, making them more effective at certain pH ranges.

3. **Synergy**: Some antioxidants act synergistically when combined, amplifying their total antioxidant capacity. For example, **vitamin C** can regenerate **vitamin E**, enhancing its antioxidant properties

5.8.1.3 Applications of Antioxidants

1. **Food Industry**: In the food industry, antioxidants are widely used to prevent **lipid oxidation**, which causes rancidity and spoils food products like oils, meats, and snacks. For example, **rosemary extract** and **vitamin E** are commonly added to oils to extend shelf life by preventing oxidative degradation [34].

2. **Pharmaceuticals**: Antioxidants are used in pharmaceuticals to prevent oxidative degradation of drugs, improve their stability, and enhance bioavailability. **Curcumin**, a compound in turmeric, is a potent antioxidant known for its anti-inflammatory and neuroprotective properties and is being studied for its therapeutic potential in Alzheimer's disease [35].

3. **Cosmetics**: Antioxidants like vitamin C, retinol, and green tea polyphenols are added to skincare products to reduce the appearance of fine lines, protect the skin from UV-induced damage, and prevent premature aging. These antioxidants neutralize free radicals generated by UV radiation, reducing oxidative stress and preventing skin damage [36].

5.8.2 Antimicrobial Activity: Mechanisms and Importance

Antimicrobial agents are substances that inhibit the growth of or kill microorganisms, including bacteria, viruses, fungi, and parasites. They are crucial in medicine for preventing infections, in food preservation, and in agriculture for controlling plant diseases. The antimicrobial activity of a substance is determined by its ability to interfere with the growth or survival of microorganisms through various mechanisms [37].

Antimicrobial agents are typically classified based on their spectrum of activity—broad-spectrum (effective against a wide range of microorganisms) or narrow-spectrum (effective against specific groups of microorganisms) [38].

5.8.2.1 Methods of Measuring Antimicrobial Activity

The minimum inhibitory concentration (MIC) is one of the most common measures of antimicrobial activity. It is the lowest concentration of an antimicrobial agent required to inhibit the growth of a microorganism. The minimum bactericidal concentration (MBC) is the concentration required to kill the microorganism [39]:

1. **Disc Diffusion Method**: This method involves placing antimicrobial-impregnated paper discs on an agar plate inoculated with microorganisms. The area around the disc where no microbial growth occurs is called the zone of inhibition, and its size is proportional to the antimicrobial activity of the substance. The zone size can vary, with antibiotics like penicillin often showing large inhibition zones (10–30 mm), while plant extracts typically show smaller zones (5–10 mm), indicating their moderate antimicrobial activity [40].
2. **Broth Dilution Method**: The MIC is determined by serially diluting an antimicrobial agent in a broth culture and observing the lowest concentration that prevents visible growth. The MIC for common antibiotics such as penicillin might range from 0.01 to 2 µg/mL, while natural plant extracts like garlic may show MICs in the range of 1–10 mg/mL, depending on the species tested [41].
3. **Time-Kill Assays**: These assays assess the bactericidal activity of an antimicrobial agent over time. For example, the time-kill assay can show that bacterial species like *E. coli* are killed within 2–4 h by agents like thymol or eugenol [42].

5.8.2.2 Mechanisms of Antimicrobial Activity

1. **Cell Wall Disruption**: Some antimicrobial agents, like penicillin and cephalosporins, inhibit the synthesis of peptidoglycan, an essential component of bacterial cell walls. This leads to the rupture of the bacterial cell due to osmotic pressure. The MIC of penicillin for *Staphylococcus aureus* typically ranges from 0.1 to 0.5 µg/mL, while ampicillin shows MIC values around 0.5–2 µg/mL for the same pathogen [43].
2. **Protein Synthesis Inhibition**: Certain antimicrobial agents, such as tetracycline, inhibit bacterial protein synthesis by binding to bacterial ribosomes, thereby pre-

venting translation. The MIC of tetracycline against *Escherichia coli* is typically around 1–2 µg/mL [44].

3. **DNA/RNA Disruption**: Some agents, such as fluoroquinolones (e.g., ciprofloxacin), inhibit DNA gyrase and topoisomerase, enzymes essential for DNA replication. The MIC of ciprofloxacin for *Salmonella* typically ranges from 0.03 to 1 µg/mL [45].

4. **Membrane Disruption**: Antimicrobial agents like polymyxin B disrupt the integrity of the bacterial cell membrane, causing leakage of cellular contents. This mechanism is particularly effective against gram-negative bacteria, which have an outer membrane that is sensitive to such agents [46].

5.8.2.3 Applications of Antimicrobial Agents

1. **Pharmaceuticals**: The use of antimicrobial agents in medicine is widespread, including antibiotics for bacterial infections, antifungals for fungal diseases, and antivirals for treating viral infections. The emergence of antibiotic resistance is a significant challenge in modern medicine, prompting the need for new antimicrobial agents and alternative therapies like phage therapy [47].

2. **Food Preservation**: Antimicrobial agents are extensively used in food preservation to prevent microbial growth and spoilage. Essential oils (e.g., oregano oil and clove oil) and natural extracts (e.g., garlic and ginger) exhibit antimicrobial properties that are harnessed in food processing to enhance the shelf life of products such as meats, dairy, and sauces [48].

3. **Agriculture**: Antimicrobials are employed to control plant diseases caused by fungi, bacteria, and viruses. Natural antimicrobial agents, including neem extract and garlic, have been researched as eco-friendly alternatives to synthetic fungicides and bactericides in organic farming [49].

References

[1]   Bahadur, S., Sahu, U. K., Sahu, D., Sahu, G., & Roy, A. (2017). Review on natural gums and mucilage and their application as excipient. *Journal of applied pharmaceutical research*, 5(4), 13–21.

[2]   Kumar, S., & Gupta, S. K. (2012). Natural polymers, gums and mucilages as excipients in drug delivery. *Polim. Med*, 42(3–4), 191–197.

[3]   Taru, P., Walunj, D., Sayd, S., & Saindane, R. (2025). Gums and Mucilages: Versatile Natural Polymers. In *Innovative Pharmaceutical Excipients: Natural Sources* (pp. 209–227). Singapore: Springer Nature Singapore.

[4]   Shiam, M. A. H., Islam, M. S., Ahmad, I., & Haque, S. S. (2025). A review of plant-derived gums and mucilages: Structural chemistry, film forming properties and application. *Journal of Plastic Film & Sheeting*, 41(2), 195–237.

[5]   Tosif, M. M., Najda, A., Bains, A., Kaushik, R., Dhull, S. B., Chawla, P., & Walasek-Janusz, M. (2021). A comprehensive review on plant-derived mucilage: characterization, functional properties, applications, and its utilization for nanocarrier fabrication. *Polymers*, *13*(7), 1066.

[6]   Haug, I. J., Draget, K. I., & Smidsrød, O. (2004). Physical and rheological properties of fish gelatin compared to mammalian gelatin. *Food hydrocolloids*, *18*(2), 203–213.

[7]   Lee, W. K., Lim, Y. Y., Leow, A. T. C., Namasivayam, P., Abdullah, J. O., & Ho, C. L. (2017). Factors affecting yield and gelling properties of agar. *Journal of Applied Phycology*, *29*(3), 1527–1540.

[8]   Urias-Orona, V., Rascón-Chu, A., Lizardi-Mendoza, J., Carvajal-Millán, E., Gardea, A. A., & Ramírez-Wong, B. (2010). A novel pectin material: extraction, characterization and gelling properties. *International journal of molecular sciences*, *11*(10), 3686–3695.

[9]   Garti, N., & Leser, M. E. (2001). Emulsification properties of hydrocolloids. *Polymers for advanced Technologies*, *12*(1-2), 123–135.

[10] Cabezas, D. M., Madoery, R., Diehl, B. W., & Tomás, M. C. (2012). Emulsifying properties of different modified sunflower lecithins. *Journal of the American Oil Chemists' Society*, *89*(2), 355–361.

[11] Moonen, H., & Bas, H. (2014). Mono-and diglycerides. Emulsifiers in food technology, 73–92.

[12] Rodríguez-López, L., Rincón-Fontán, M., Vecino, X., Cruz, J. M., & Moldes, A. B. (2018). Biological surfactants vs. polysorbates: Comparison of their emulsifier and surfactant properties. *Tenside Surfactants Detergents*, *55*(4), 273–280.

[13] Koocheki, A., Mortazavi, S. A., Shahidi, F., Razavi, S. M. A., & Taherian, A. R. (2009). Rheological properties of mucilage extracted from Alyssum homolocarpum seed as a new source of thickening agent. *Journal of food engineering*, *91*(3), 490–496.

[14] Mandala, I. G. (2012). Viscoelastic properties of starch and non-starch thickeners in simple mixtures or model food. *Viscoelasticity: From Theory to Biological Applications. InTech. England. pp*, 217–236.

[15] Urlacher, B., & Dalbe, B. (1997). Xanthan gum. In *Thickening and gelling agents for food* (pp. 202–226). Boston, MA: Springer US.

[16] Mudgil, D., Barak, S., & Khatkar, B. S. (2014). Guar gum: processing, properties and food applications—a review. *Journal of food science and technology*, *51*(3), 409–418.

[17] Thomas, W. R. (1997). Carrageenan. In *Thickening and gelling agents for food* (pp. 45–59). Boston, MA: Springer US.

[18] Pérez-Orozco, J. P., Sánchez-Herrera, L. M., & Ortiz-Basurto, R. I. (2019). Effect of concentration, temperature, pH, co-solutes on the rheological properties of Hyptis suaveolens L. mucilage dispersions. *Food Hydrocolloids*, *87*, 297–306.

[19] Medina-Torres, L., Brito-De La Fuente, E., Torrestiana-Sanchez, B., & Katthain, R. (2000). Rheological properties of the mucilage gum (Opuntia ficus indica). *Food hydrocolloids*, *14*(5), 417–424.

[20] Sangwan, Y. S., Sngwan, S., Jalwal, P., Murti, K., & Kaushik, M. (2011). Mucilages and their pharmaceutical applications: an overview. *Pharmacologyonline*, *2*, 1265–1271.

[21] Sierra-López, L. D., Hernandez-Tenorio, F., Marín-Palacio, L. D., & Giraldo-Estrada, C. (2023). Coffee mucilage clarification: A promising raw material for the food industry. *Food and Humanity*, *1*, 689–695.

[22] Martins, V. B., Da Silva Carvalho, J. G., PIETRO, B., GABRIELLI, A., ALVES DA CUNHA, M. A., KLEIN DAS NEVES, J. C., . . . & BUDZIAK PARABOCZ, C. R. (2021). Taro Mucilage: Extraction, Characterization, and Application in Cosmetic Formulations. *Journal of Cosmetic Science*, May 1;72(3).

[23] Medina-Torres, L., Brito-De La Fuente, E., Torrestiana-Sanchez, B., & Katthain, R. (2000). Rheological properties of the mucilage gum (Opuntia ficus indica). *Food hydrocolloids*, *14*(5), 417–424.

[24] Soukoulis, C., Gaiani, C., & Hoffmann, L. (2018). Plant seed mucilage as emerging biopolymer in food industry applications. *Current Opinion in Food Science*, *22*, 28–42.

[25] Sangwan, Y. S., Sngwan, S., Jalwal, P., Murti, K., & Kaushik, M. (2011). Mucilages and their pharmaceutical applications: an overview. *Pharmacologyonline*, *2*, 1265–1271.

[26]   Liu, M., Gan, Z., Jia, B., Hou, Y., Zheng, H., Wu, Y., . . . & Guo, Z. (2022). Mucilage-inspired robust antifouling coatings under liquid mediums. *Chemical Engineering Journal, 446*, 136949.
[27]   Muñoz, L. A., Cobos, A., Diaz, O., & Aguilera, J. M. (2012). Chia seeds: Microstructure, mucilage extraction and hydration. *Journal of food Engineering, 108*(1), 216–224.
[28]   Kaçar, D. (2008). *Screening of some plant species for their total antioxidant and antimicrobial activities* (Master's thesis, Izmir Institute of Technology (Turkey)).
[29]   Santos-Sánchez, N. F., Salas-Coronado, R., Villanueva-Cañongo, C., & Hernández-Carlos, B. (2019). *Antioxidant compounds and their antioxidant mechanism*. IntechOpen.
[30]   Dawidowicz, A. L., Wianowska, D., & Olszowy, M. (2012). On practical problems in estimation of antioxidant activity of compounds by DPPH method (Problems in estimation of antioxidant activity). *Food chemistry, 131*(3), 1037–1043.
[31]   Ozgen, M., Reese, R. N., Tulio, A. Z., Scheerens, J. C., & Miller, A. R. (2006). Modified 2, 2-azino-bis-3-ethylbenzothiazoline-6-sulfonic acid (ABTS) method to measure antioxidant capacity of selected small fruits and comparison to ferric reducing antioxidant power (FRAP) and 2, 2 '-diphenyl-1-picrylhydrazyl (DPPH) methods. *Journal of agricultural and food chemistry, 54*(4), 1151–1157.
[32]   Gohari, A. R., Hajimehdipoor, H., Saeidnia, S., Ajani, Y., & Hadjiakhoondi, A. (2011). Antioxidant activity of some medicinal species using FRAP assay (2011): 54–60.
[33]   Zulueta, A., Esteve, M. J., & Frígola, A. (2009). ORAC and TEAC assays comparison to measure the antioxidant capacity of food products. *Food chemistry, 114*(1), 310–316.
[34]   Pokorny, J., Yanishlieva, N., & Gordon, M. H. (Eds.). (2001). *Antioxidants in food: practical applications*. CRC press.
[35]   Losada-Barreiro, S., Sezgin-Bayindir, Z., Paiva-Martins, F., & Bravo-Díaz, C. (2022). Biochemistry of antioxidants: Mechanisms and pharmaceutical applications. *Biomedicines, 10*(12), 3051.
[36]   de Lima Cherubim, D. J., Buzanello Martins, C. V., Oliveira Fariña, L., & da Silva de Lucca, R. A. (2020). Polyphenols as natural antioxidants in cosmetics applications. *Journal of cosmetic dermatology, 19*(1), 33–37.
[37]   Russell, A. D. (1991). Principles of antimicrobial activity. *Disinfection, sterilization and preservation, 3*, 717–745.
[38]   Lemire, J. A., Harrison, J. J., & Turner, R. J. (2013). Antimicrobial activity of metals: mechanisms, molecular targets and applications. *Nature Reviews Microbiology, 11*(6), 371–384.
[39]   Balouiri, M., Sadiki, M., & Ibnsouda, S. K. (2016). Methods for in vitro evaluating antimicrobial activity: A review. *Journal of pharmaceutical analysis, 6*(2), 71–79.
[40]   Zaidan, M. R., Noor Rain, A., Badrul, A. R., Adlin, A., Norazah, A., & Zakiah, I. (2005). In vitro screening of five local medicinal plants for antibacterial activity using disc diffusion method. *Trop biomed, 22*(2), 165–170.
[41]   Wiegand, I., Hilpert, K., & Hancock, R. E. (2008). Agar and broth dilution methods to determine the minimal inhibitory concentration (MIC) of antimicrobial substances. *Nature protocols, 3*(2), 163–175.
[42]   Li, F., Weir, M. D., Fouad, A. F., & Xu, H. H. (2013). Time-kill behaviour against eight bacterial species and cytotoxicity of antibacterial monomers. *Journal of dentistry, 41*(10), 881–891.
[43]   Rurián-Henares, J. A., & Morales, F. J. (2008). Antimicrobial activity of melanoidins against Escherichia coli is mediated by a membrane-damage mechanism. *Journal of Agricultural and Food Chemistry, 56*(7), 2357–2362.
[44]   Anandabaskar, N. (2021). Protein synthesis inhibitors. In *Introduction to basics of pharmacology and toxicology: volume 2: essentials of systemic pharmacology: from principles to practice* (pp. 835–868). Singapore: Springer Nature Singapore.
[45]   Gagandeep, K. R., Narasingappa, R. B., & Vyas, G. V. (2024). Unveiling mechanisms of antimicrobial peptide: Actions beyond the membranes disruption. *Heliyon*, Oct 15;10(19).
[46]   Benfield, A. H., & Henriques, S. T. (2020). Mode-of-action of antimicrobial peptides: membrane disruption vs intracellular mechanisms. *Frontiers in Medical Technology, 2*, 610997.

[47]   Russell, A. D., Jenkins, J., & Harrison, I. H. (1968). The inclusion of antimicrobial agents in
       pharmaceutical products. *Advances in Applied Microbiology, 9*, 1–38.
[48]   Tiwari, B. K., Valdramidis, V. P., O'Donnell, C. P., Muthukumarappan, K., Bourke, P., & Cullen,
       P. J. (2009). Application of natural antimicrobials for food preservation. *Journal of agricultural and
       food chemistry, 57*(14), 5987–6000.
[49]   Keymanesh, K., Soltani, S., & Sardari, S. (2009). Application of antimicrobial peptides in agriculture
       and food industry. *World Journal of Microbiology and Biotechnology, 25*(6), 933–944.

6 Environmental Impact and Biodegradability of Natural Polymers

6.1 Introduction

Synthetic polymers, also known as plastics, have become an integral part of modern life due to their versatility, cost-effectiveness, and wide range of applications [1]. From packaging materials to medical devices, and automotive parts to household goods, synthetic polymers are found in nearly every sector of industry and daily life [2, 3]. However, the very properties that make these materials so useful have also led to significant environmental concerns. The rapid proliferation of synthetic polymers, particularly in single-use applications, has created a massive global waste problem, mentioned in Figure 6.1.

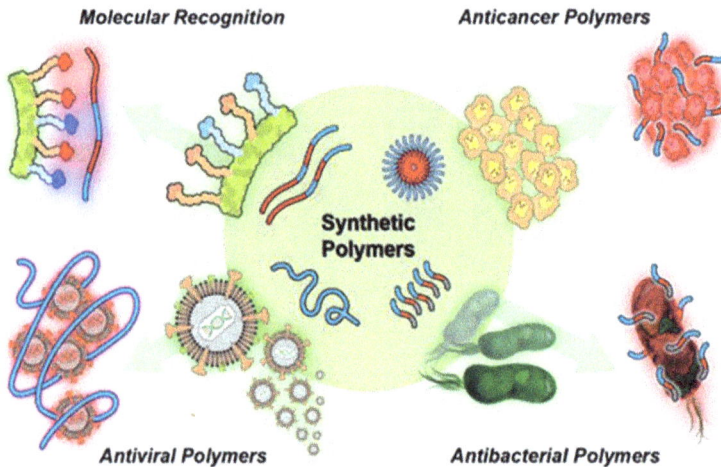

Figure 6.1: Environmental factors of synthetic polymers. Reproduced with permission from [4].

The environmental impact of synthetic polymers begins with their production. Most synthetic polymers are derived from petrochemical feedstocks, primarily from nonrenewable fossil fuels like oil and natural gas [1]. The extraction, refinement, and processing of these materials contribute significantly to greenhouse gas emissions, air pollution, and the depletion of natural resources. Furthermore, the energy-intensive production process, along with the use of toxic chemicals in the synthesis of some polymers, exacerbates the environmental footprint of plastics [3]. Once produced, synthetic polymers are often used in single-use products such as plastic bottles, food packaging, straws, and bags that are discarded after a short period of use [4, 5]. The sheer volume of plastic waste generated globally, estimated to be over 300 million

https://doi.org/10.1515/9783111673509-006

tons annually, has overwhelmed waste management systems and led to the accumulation of plastics in landfills, rivers, and oceans.

One of the most pressing environmental concerns associated with synthetic polymers is plastic pollution. Large quantities of plastic waste end up in the oceans, where they can remain for centuries, breaking into smaller particles known as microplastics [4, 5]. These tiny fragments are not only unsightly but also hazardous to marine life. Many marine animals mistake plastic for food, ingesting it either directly or through the consumption of smaller organisms that have consumed plastic particles [6]. The ingestion of plastics can cause physical harm, blockages in the digestive system, and even death. Moreover, plastics can leach harmful chemicals into the water, including additives used during production (such as plasticizers and flame retardants), which can accumulate in the tissues of marine organisms and enter the food chain. As a result, plastics pose significant risks to biodiversity and ecosystem health, with far-reaching consequences for human populations that rely on marine resources for food and livelihoods [4, 5, 7].

In addition to their impact on aquatic ecosystems, synthetic polymers also contribute to terrestrial pollution. Plastics often find their way into landfills, where they can take hundreds of years to degrade, occupying valuable space and potentially releasing toxic substances into the soil and groundwater [8]. While some types of plastics like polyethylene and polypropylene are relatively stable, others such as polyvinyl chloride (PVC) contain additives and chemical compounds that can be hazardous to the environment. Furthermore, the persistence of plastics in landfills creates a long-term waste management challenge, with few viable solutions for large-scale plastic waste disposal. Recycling programs for synthetic polymers have been implemented in many regions, but these efforts have been hindered by logistical challenges, the contamination of recyclables, and the complexity of sorting different types of plastics [7, 8]. As a result, the majority of plastic waste is either incinerated, which releases harmful pollutants into the air, ends up in landfills or the natural environment [1].

Another environmental issue associated with synthetic polymers is the impact on human health. Many plastics contain chemicals known to be endocrine disruptors, such as bisphenol A (BPA), phthalates, and various flame retardants [1, 2]. These chemicals can leach out of plastic products over time and enter the human body through ingestion, inhalation, or skin contact. Studies have shown that exposure to such chemicals can interfere with hormone systems, potentially leading to developmental, reproductive, and neurological problems. BPA, in particular, has been linked to a range of health issues, including increased risk of certain cancers, cardiovascular disease, and obesity. Moreover, the widespread use of plastics in food packaging has raised concerns about the migration of harmful substances from plastic containers into food and beverages, especially when plastics are exposed to heat or acidic conditions. The accumulation of microplastics in the environment also presents a growing concern for human health, as these tiny particles have been found in drinking water, seafood, and even air [8].

The environmental and health impacts of synthetic polymers have spurred efforts to develop alternatives to plastic products, as well as to improve plastic waste management systems. Biodegradable plastics, made from renewable resources such as corn starch, are seen as a potential solution to some of the problems associated with conventional plastics. However, the environmental benefits of biodegradable plastics are still debated, as they may require specific conditions to break down and may not fully degrade in natural environments. Furthermore, the production of biodegradable plastics still requires energy and resources, and their widespread adoption could divert land from food production. Another promising area of research is the development of "green" polymers made from renewable materials such as algae, cellulose, or plant-based oils [9]. These materials offer the potential for more sustainable plastic alternatives, although their production and scalability remain significant challenges.

On the waste management front, increased public awareness and policy initiatives are helping to address plastic pollution [4]. Many countries and cities have introduced bans or restrictions on single-use plastics, such as plastic bags, straws, and cutlery, in an effort to reduce plastic waste and encourage the use of reusable alternatives [5]. Recycling infrastructure has also been improved in some regions, and advances in chemical recycling technologies offer the potential to break down plastics into their constituent monomers for reuse. However, global recycling rates for plastics remain low, and the demand for recycled plastic materials often exceeds the supply. As a result, the most effective solution to the plastic pollution crisis will likely require a combination of approaches, including reducing plastic consumption, increasing recycling rates, and investing in the development of alternative materials [1, 9].

6.2 Biodegradability of Mucilage and Gums

Mucilages and gums, derived primarily from plant sources, have gained considerable attention due to their biodegradable nature, positioning them as more environmentally friendly alternatives to synthetic polymers [10]. These natural polysaccharides are found in various plant species, serving vital functions like water retention, seed protection, and as defense mechanisms against herbivores. The widespread use of synthetic polymers in everyday products has led to increasing concerns about their environmental impact, particularly regarding their persistence in the environment and slow degradation rates [4]. This has placed mucilages and gums under scrutiny as biodegradable materials that can break down more rapidly and safely, offering a promising alternative in various industries, such as food, pharmaceuticals, and cosmetics [11].

The biodegradability of mucilage and gums refers to the process by which these natural substances are decomposed by microorganisms such as bacteria, fungi, and enzymes into simpler, nontoxic by-products [10]. This process is primarily driven by the enzymatic breakdown of polysaccharide chains into simpler sugar units, which microorganisms can utilize as a source of energy. The time required for complete bio-degradation varies, influenced by the chemical structure of the polysaccharide, environmental conditions like temperature and humidity, and the presence of microorganisms capable of breaking down these substances. Unlike synthetic polymers, which can remain in the environment for hundreds of years, mucilages and gums are relatively short-lived, providing a much lower environmental footprint [12].

One key aspect of mucilage and gum biodegradability is the structure of the polysaccharides themselves. These substances are often made of complex sugars, such as glucose, galactose, and mannose, which are linked in varying patterns. While mucilages are typically simpler structures that are easier for microorganisms to degrade, gums tend to have more complex molecular structures, including protein and uronic acid groups, which can slow down the degradation process. However, even though gums like gum arabic or guar gum may have more intricate structures, they are generally still biodegrade, much faster than synthetic polymers [13]. Research indicates that most gums break down within 30–60 days in natural environments, depending on the conditions and microbial populations present. For comparison, the degradation of common synthetic polymers such as polyethylene or polypropylene can take anywhere from 500 to 1,000 years. This stark difference underscores the environmental advantages of mucilages and gums in terms of biodegradability [13, 14].

Mucilages and gums, when introduced into various environmental conditions, undergo biodegradation, primarily through microbial activity. Microorganisms, including bacteria, fungi, and actinomycetes, produce enzymes such as cellulase, pectinase, and amylase that target and break down the polysaccharides into simpler molecules like monosaccharides (glucose, fructose) and oligosaccharides [9, 10]. These simple sugars are then metabolized by microorganisms, contributing to the natural nutrient cycle. Studies have demonstrated that the presence of microorganisms in soil, water, or even in wastewater treatment facilities significantly accelerates the breakdown of mucilages and gums [12]. For example, research on guar gum has shown that it degrades within 30 days in soils with active microbial communities, where it serves as a carbon source for microorganisms. Similarly, in aquatic environments, the biodegradation of gum arabic and other polysaccharides typically occurs within 30–60 days, depending on the water temperature, microbial activity, and presence of oxygen.

One critical factor influencing the biodegradation of mucilages and gums is environmental conditions, particularly temperature, moisture, and the abundance of microorganisms. In humid and warm climates, biodegradation is typically faster, with temperatures above 20 °C enhancing microbial activity [13]. Under such conditions, microbial breakdown can lead to the near-complete degradation of the material within a few weeks. For example, mucilage from okra or flaxseed, which is highly hydrophilic and soluble in water, has been shown to degrade within 15–30 days when exposed to warm and moist environment. On the other hand, in colder or drier conditions, the degradation rate significantly slows down, and the substances may persist for longer periods, though they eventually degrade due to their natural composition [14].

While mucilage and gums are generally biodegradable, the rate of degradation can vary based on the structure and complexity of the polysaccharide. Simple mucilages, such as those derived from flax or chia seeds, are more readily biodegraded due to their linear sugar chains and the lack of complex branching structures [15]. In contrast, gums like acacia or tragacanth, which contain protein and uronic acid side chains, may take longer to degrade because these side chains require additional enzymatic actions [16]. For instance, gum arabic, a widely used polysaccharide, has a more complex structure that slows its degradation. However, even complex gums are still considered highly biodegradable compared to synthetic alternatives. On average, the degradation time for these substances in natural conditions ranges from 30 to 60 days, with environmental factors playing a crucial role in influencing this time-frame [13, 14].

The biodegradability of mucilages and gums also has implications for soil and water quality. In agricultural settings, where these substances are used as thickeners, stabilizers, or soil conditioners, their biodegradation contributes positively to soil health [17]. The breakdown products of mucilages and gums, such as glucose and other saccharides, provide an additional carbon source for soil microorganisms, thus promoting microbial diversity and activity. This process enhances soil fertility, improving water retention and nutrient availability. Furthermore, mucilages and gums have been shown to have a positive impact on the aggregation of soil particles, helping prevent erosion and improving soil structure [17]. In aquatic ecosystems, the degradation of gums and mucilages can similarly improve water quality by promoting the breakdown of organic matter and reducing the potential for eutrophication.

Despite these advantages, the use of mucilages and gums must be carefully managed to avoid potential negative environmental impacts. For instance, the release of large quantities of gums and mucilages into wastewater treatment systems or natural water bodies may overwhelm microbial communities, leading to an imbalance in the ecosystem. In wastewater treatment plants, excessive concentrations of mucilage and gum residues can lead to clogging of filters and pipes, slowing down the treatment process [18]. However, such issues are generally less severe compared to the impact of synthetic polymers, which do not degrade at all under typical environmental conditions. Proper management of mucilage and gum waste, through recycling and appropriate disposal methods, can mitigate these risks and ensure that these materials contribute positively to the environment [19].

The renewable nature of mucilages and gums further enhances their biodegradability profile. Unlike synthetic plastics, which are derived from finite petroleum resources, mucilages and gums come from plants that can be sustainably cultivated year after year [13]. This makes them an attractive option for industries looking to reduce their environmental footprint. The cultivation of plants producing these polysaccharides can also provide economic benefits to farmers and rural communities, promoting sustainable agriculture practices [17]. For example, the global market for guar gum, primarily produced in India and Pakistan, has seen significant growth, offering income opportunities for local farmers. Additionally, the relatively low environmental cost of producing mucilages and gums, compared to petroleum-based plastics, further strengthens their role as sustainable alternatives in industries that prioritize biodegradability.

6.3 Life Cycle Analysis (LCA) of Natural Polymers

Natural polymers include materials like starch, cellulose, chitosan, proteins (such as casein), and biopolymers like polylactic acid (PLA), which are derived from renewable sources such as plants, animals, and microorganisms. These polymers are generally biodegradable, and many are nontoxic, offering an environmentally friendly alternative to synthetic polymers, such as polyethylene and polypropylene [20].

6.3.1 Goal and Scope Definition

In this first phase of life cycle analysis (LCA), the purpose of the analysis is defined, and the system boundaries are set. For natural polymers, this would typically involve defining the life cycle from raw material extraction (e.g., harvesting of plant or animal-based materials), processing (such as conversion into polymers), product use, and disposal or end-of-life (EOL) management [21]. The system boundaries help determine which processes are included in the analysis (e.g., cultivation, transportation, production, recycling, and disposal) and what environmental impact categories will be assessed, such as carbon footprint, water use, land use, and toxicity [22].

6.3.2 Inventory Analysis (Life Cycle Inventory – LCI)

In this stage, all inputs (resources, energy, materials) and outputs (emissions, waste, by-products) of each life cycle stage are quantified. For natural polymers, this would include:

- **Raw Material Extraction**: This stage includes the cultivation or harvesting of biomass, such as corn for PLA or cotton for cellulose-based products. Inputs such as water, fertilizers, and pesticides, as well as the associated environmental impacts (e.g., land use, water use, and emissions from agricultural practices) are tracked.
- **Processing**: This involves the conversion of raw materials into usable natural polymers. Energy consumption during production, water usage, waste products

(such as by-products or emissions), and chemicals used in polymer extraction or modification are accounted for.

- **Product Use**: The LCA considers the energy consumption, resource usage, and any other environmental impacts that occur during the use phase of the product made from the natural polymer. For instance, PLA used in packaging may involve energy usage during food storage or transportation.
- **EOL Management**: This phase includes the disposal of the product after its use, such as composting, recycling, incineration, or landfill. For natural polymers, biodegradability plays a significant role here. A key aspect of the analysis is to track how the natural polymer breaks down and the environmental impact of its degradation, whether it involves microbial decomposition in composting or more hazardous processes like incineration [21].

6.3.3 Impact Assessment

The impact assessment phase evaluates the potential environmental impacts identified during the inventory analysis. Several categories of impact are typically considered in an LCA of natural polymers:

- **Global Warming Potential (GWP)**: GWP is measured as the total carbon dioxide-equivalent emissions produced during the entire life cycle. While natural polymers are derived from renewable sources, the cultivation and processing stages can still produce greenhouse gases (GHGs), especially if energy-intensive practices are used. For instance, the production of PLA from corn involves energy usage and agricultural emissions that can contribute to GWP.
- **Water Use**: The water consumption involved in growing raw materials, such as the vast amounts of water needed for cotton or corn production, can be significant. LCA tracks the water footprint to assess whether the production of natural polymers is sustainable, particularly in regions facing water scarcity.
- **Land Use**: Cultivating crops for natural polymers requires large areas of land, which can lead to land-use changes, such as deforestation, monoculture farming, or soil degradation. An LCA of natural polymers would account for the land area required for raw material cultivation and the environmental consequences of large-scale agricultural production.
- **Energy Use**: The energy consumed in extracting raw materials, processing them into polymer products, and transporting them to markets plays a crucial role in assessing the overall environmental performance of natural polymers. Renewable energy use can mitigate some of the impacts, but if fossil fuels are used extensively during these stages, the environmental benefits may be reduced.
- **Waste and EOL**: This category assesses the fate of the polymer after it is used. Biodegradable polymers such as PLA and PHA (polyhydroxyalkanoates) typically break down in natural environments, reducing long-term pollution. However, if

they are improperly disposed of or incinerated, they could still contribute to environmental degradation [22].

6.3.4 Interpretation

The interpretation phase of the LCA involves analyzing the results from the previous stages to identify opportunities for improving the sustainability of natural polymers. For instance, if the production of a particular natural polymer shows high water consumption or large greenhouse gas emissions during its cultivation phase, this insight may drive changes in agricultural practices or product design. Additionally, identifying the most environmentally damaging stages allows manufacturers to prioritize changes that would have the greatest environmental benefit, such as transitioning to renewable energy sources for production or improving EOL management strategies like increasing composting infrastructure.

6.4 Environmental Benefits of Natural Polymers

6.4.1 Lower Carbon Footprint

Natural polymers generally have a lower carbon footprint than synthetic alternatives because they are derived from renewable resources. For example, PLA, a biodegradable plastic made from fermented plant sugars, has a significantly lower carbon footprint during production compared to traditional plastics like polyethylene and polypropylene. This is due to the renewable nature of the raw materials and the potential for carbon sequestration during plant growth. While PLA still involves energy and water use during cultivation and processing, its carbon emissions are much lower than those from fossil fuel-based polymers [23].

6.4.2 Biodegradability and Reduced Pollution

One of the most significant environmental advantages of natural polymers is their biodegradability. Unlike synthetic plastics, which persist in the environment for hundreds of years, natural polymers typically decompose through microbial activity. This makes them less harmful to ecosystems and wildlife. For instance, cellulose-based materials like paper and cotton fibers break down in composting environments, reducing landfill waste and pollution. However, the biodegradability of natural polymers can vary based on environmental conditions, and factors such as soil composition and temperature, while moisture levels may influence the degradation rate [24].

6.4.3 Renewable Resource Use

Natural polymers are made from renewable biological resources, such as crops or agricultural residues, which can be replenished every year. This contrasts with synthetic polymers, which rely on finite petroleum reserves. As agriculture practices become more sustainable, the environmental impact of cultivating natural polymers can be reduced, making them a more viable long-term alternative to petrochemical-based plastics [25].

6.5 Natural Polymers in Waste Management

The use of natural polymers in waste management revolves around their ability to reduce the volume of nonbiodegradable waste, facilitate composting, and enable resource recovery. Moreover, these polymers can also enhance waste treatment processes such as improving the efficiency of landfill operations, wastewater treatment, and recycling systems. This ability to interact favorably with the environment makes them crucial in the transition toward a circular economy, where products are designed for reuse, recycling, and minimal waste generation [18, 25].

6.5.1 Key Types of Natural Polymers in Waste Management

– **Starch-Based Polymers**
Starch is one of the most widely used natural polymers in waste management. It is biodegradable, abundant, and relatively inexpensive, making it a prime candidate for various applications. Starch-based polymers are used in food packaging, biodegradable bags, and agricultural films. These products can be composted at the end of their life cycle, reducing the need for landfill space and minimizing the environmental impact of plastic waste. In fact, starch-based films typically degrade within 1 to 2 months when exposed to natural conditions, providing a substantial reduction in waste accumulation compared to synthetic plastics. Starch is typically used in combination with other biodegradable polymers such as polyvinyl alcohol (PVA) or PLA to improve its mechanical properties and processing capabilities. The advantage of starch-based biopolymers is that they not only degrade more quickly but also return valuable organic matter to the soil when composted, enhancing soil fertility [26].

– **Cellulose-Based Polymers**
Cellulose, the most abundant natural polymer on Earth, is a major component of plant cell walls. It is widely used in waste management for producing biodegradable films, coatings, and packaging materials. Cellulose-based polymers can be derived from renewable sources such as wood, cotton, and agricultural waste. The biodegra-

dation of cellulose is rapid, and it typically breaks down within 2 to 6 months, depending on environmental conditions. Cellulose can be chemically modified to form various derivatives such as carboxymethyl cellulose (CMC) and hydroxypropyl cellulose (HPC), which are used in the production of biodegradable plastics. These cellulose derivatives are particularly useful in the production of packaging films, sanitary products, and medical waste management items, where their biodegradability is a significant benefit in reducing landfill waste. Additionally, cellulose-based bioplastics can be composted, further reducing environmental impacts [27].

– PLA

PLA is one of the most well-known biodegradable polymers and is produced from renewable resources such as corn starch or sugarcane. PLA has been widely used in food packaging, disposable tableware, and agricultural films. PLA biodegrades in industrial composting environments within 60 to 180 days, depending on the temperature and moisture conditions. However, PLA's biodegradability is slower in home composting environments, and improper disposal, such as landfill burial, can slow down its degradation. PLA has a significant advantage over conventional plastics because it does not release harmful chemicals during degradation. As it is made from renewable resources, PLA helps to reduce reliance on fossil fuels and carbon emissions. However, there are concerns regarding the land-use implications of large-scale corn cultivation, which can lead to deforestation, pesticide use, and water consumption. Nonetheless, when managed properly, PLA can offer a sustainable alternative to conventional plastics in waste management [28].

– Polyhydroxyalkanoates (PHA)

Polyhydroxyalkanoates (PHA) are a family of biodegradable plastics produced by microorganisms, through the fermentation of renewable resources such as sugars and vegetable oils. PHAs are biodegradable in both aerobic and anaerobic conditions, making them particularly useful in waste management applications such as packaging, agricultural films, and biomedical products. PHA-based plastics decompose within weeks to months in natural environments, leaving no toxic residues. The production of PHA is still relatively expensive compared to conventional plastics, but ongoing research and technological advancements are making PHA more cost competitive. Additionally, PHAs can be produced from waste materials like food scraps or agricultural waste, helping reduce both plastic pollution and organic waste. PHA's high biodegradability and low environmental impact make it a strong contender for sustainable waste management practices [29].

– Chitosan

Chitosan is a natural polymer derived from the shells of crustaceans, such as shrimp and crabs. It is biodegradable, nontoxic, and has antimicrobial properties, which makes it suitable for a variety of waste management applications. Chitosan is often used in wastewater treatment, where it acts as a flocculant to remove heavy metals, oils, and other pollutants. Additionally, it is employed in the production of biodegradable films and coatings for packaging, medical applications, and food waste management. Chitosan's biodegradability and ability to form gels and films also make it useful in composting and waste treatment processes, as it enhances the breakdown of organic waste materials. Moreover, its low toxicity and antimicrobial properties help prevent the spread of pathogens, contributing to cleaner waste management practices [30].

– **Alginate**

Alginate is a natural polymer extracted from brown seaweed and is widely used in the food, pharmaceutical, and waste management industries. Alginate can form biodegradable films and coatings, and it is often used in the treatment of wastewater to remove pollutants. Alginate-based materials are highly biodegradable, typically breaking down within 1 to 2 months under natural conditions. Alginate is also used in the production of biodegradable packaging and agricultural films, which reduce the accumulation of plastic waste [31].

6.5.2 Role of Natural Polymers in Waste Management Systems

Natural polymers contribute to waste management in various ways, primarily through their biodegradability and potential for resource recovery. These materials can help reduce plastic waste in landfills, enhance composting processes, and facilitate more sustainable waste treatment practices.

– **Reduction of Plastic Waste in Landfills**

The most significant contribution of natural polymers to waste management is their potential to reduce the volume of plastic waste that accumulates in landfills. Synthetic plastics, such as polyethylene and polystyrene, can take hundreds of years to degrade, leading to environmental pollution, especially in marine ecosystems. In contrast, natural polymers, which break down more rapidly, help minimize landfill burden. The use of biodegradable plastics in packaging, disposable items, and other single-use products can significantly reduce the persistence of plastic waste. For instance, the widespread adoption of starch-based bags, PLA packaging, and cellulose-based products, in place of traditional plastics, can help reduce the environmental burden associated with single-use plastics. Studies have shown that replacing just 10% of conventional plastic packaging with biodegradable alternatives could reduce plastic waste by over 2 million tons annually [32].

– **Composting and Organic Waste Management**

Natural polymers play a vital role in composting systems by providing biodegradable materials that can break down alongside organic waste. For example, starch-based bioplastics and PLA are often used in compostable food packaging, cutlery, and food waste bags. These products are designed to break down in composting facilities, returning valuable organic matter to the soil. Composting not only diverts waste from landfills but also produces nutrient-rich compost that can be used to improve soil quality and promote sustainable agriculture [33].

– **Wastewater Treatment and Pollution Control**

Certain natural polymers, such as chitosan and alginate, are used in wastewater treatment to remove pollutants like heavy metals, oils, and particulate matter. Chitosan, in particular, has been shown to effectively flocculate suspended solids in wastewater, facilitating their removal [34]. This capability enhances the efficiency of wastewater treatment plants, reducing the environmental impact of industrial effluents and municipal waste. Moreover, chitosan and alginate have been used in bioremediation processes to treat polluted environments. For example, chitosan-based materials can be used to clean up oil spills, heavy metal contamination, and agricultural runoff, reducing the negative effects of industrial and agricultural waste on ecosystems [35].

– **Sustainable Packaging Solutions**

Natural polymers provide a sustainable alternative to petroleum-based plastics in packaging. By using biodegradable materials like PLA, starch-based films, and cellulose, the packaging industry can significantly reduce plastic waste. These materials can be composted or recycled, reducing the burden on waste management systems and mitigating environmental pollution. The use of natural polymers in food packaging, agricultural products, and medical supplies is gaining traction, with more manufacturers embracing eco-friendly alternatives to traditional plastics [36].

6.6 Toxicity and Eco-Safety of Mucilage and Gums

The toxicity of mucilages and gums largely depends on their botanical source, chemical composition, and concentration in use. Generally, mucilages and gums are considered safe for human consumption, and the vast majority of commercially used mucilages and gums are derived from edible plants like *Guar gum* (*Cyamopsis tetragonoloba*), *Aloe vera*, *Okra* (*Abelmoschus esculentus*), and *Acacia gum* (gum arabic). However, certain species may produce gums or mucilages that contain toxic secondary metabolites, potentially causing adverse effects when consumed in large quantities [37]:

–   **Guar Gum**: Guar gum is one of the most widely used natural gums, derived from the seeds of *Cyamopsis tetragonoloba*. It is often utilized in the food and pharmaceutical industries as a thickening, gelling, and emulsifying agent. Research indicates that guar gum is generally nontoxic at normal consumption levels, with no significant adverse effects when consumed in food quantities. However, when consumed in excess, guar gum can cause gastrointestinal issues such as bloating, gas, and diarrhea due to its high fiber content. Additionally, unprocessed guar gum may contain traces of toxic compounds like *cyanogenic glycosides*, which could release cyanide when metabolized, though this is rare and usually mitigated through processing [38].

–   **Aloe Vera Gel**: Aloe vera gel is renowned for its soothing and moisturizing properties, making it popular in cosmetics and topical applications. While aloe gel itself is safe for most people when used externally, ingestion of aloe latex (a yellowish fluid found beneath the skin of the plant) can be toxic. Aloe latex contains compounds known as *anthraquinone glycosides*, which have a potent laxative effect. Long-term or excessive consumption of aloe latex can lead to dehydration, diarrhea, and abdominal cramps. It has also been linked to potential kidney damage when taken in high doses over extended periods. As a result, aloe vera products are often processed to remove the latex, leaving only the gel, which is safer for internal use [39].

–   **Okra Mucilage**: Okra mucilage is a gel-like substance extracted from the fruit of *Abelmoschus esculentus*, a plant commonly consumed in food and used in various medicinal formulations. Okra mucilage is generally regarded as safe, with studies showing no significant toxicity at typical dietary levels. Okra is often used for its dietary fiber and mucilaginous properties, which help in regulating digestion. However, like other mucilages, excessive consumption could cause gastrointestinal discomfort such as bloating, cramps, and flatulence, particularly in individuals with sensitive digestive systems [40].

–   **Acacia Gum**: Acacia gum is derived from the sap of *Acacia* species, primarily *Acacia senegal* and *Acacia seyal*. It is widely used as a stabilizer, emulsifier, and thickening agent in food products. Gum arabic is generally recognized as safe (GRAS) by regulatory bodies such as the U.S. Food and Drug Administration (FDA). It is nontoxic and well-tolerated, even at higher concentrations. However, certain individuals may experience mild gastrointestinal symptoms like bloating or indigestion when consuming large amounts of gum arabic, especially if they are not accustomed to high-fiber foods [41].

While most of these substances are considered safe when used appropriately, toxicity concerns arise primarily from improper processing, overconsumption, or individual sensitivities. Additionally, certain mucilages and gums may cause allergic reactions in sensitive individuals, leading to symptoms such as skin rashes, itching, or respiratory distress. It is important to assess the specific origin and quality of these substances before use, ensuring they are free from contaminants or potentially harmful substances.

6.7 Eco-Safety and Environmental Impact

In addition to concerns about human safety, the ecological impact of mucilages and gums must also be carefully examined. The environmental footprint of harvesting these substances varies depending on the plant species, the extraction method, and the scale of production:

- **Sustainable Harvesting of Plant Sources**: Many mucilage- and gum-producing plants are harvested from wild or semi-wild environments. Unsustainable harvesting practices can lead to habitat destruction, overexploitation of natural resources, and the decline of native plant populations. For example, *Acacia senegal* trees, which produce gum arabic, are primarily sourced from the arid and semi-arid regions of Africa. If gum arabic is harvested without proper management, it can lead to the depletion of tree populations and disrupt local ecosystems. Sustainable agricultural practices, such as agroforestry and controlled harvesting, are essential to minimize the ecological impact of these industries [42].
- **Water Usage**: Many plants that produce mucilages and gums, such as okra and guar, require substantial amounts of water for cultivation. In regions where water resources are limited, intensive cultivation of these crops can contribute to water scarcity and stress on local water systems. For instance, guar cultivation in arid regions of India and Pakistan places significant pressure on water resources, particularly in areas where irrigation is necessary for optimal yields. Sustainable water management practices, including the use of rainwater harvesting, drip irrigation, and efficient water use technologies, are critical to mitigating these issues [43].
- **Biodiversity and Soil Health**: The large-scale cultivation of mucilage- and gum-producing crops can impact biodiversity and soil health. Monoculture farming practices, which focus on growing a single crop, can lead to a reduction in soil fertility, increased susceptibility to pests, and a decline in biodiversity. In contrast, polyculture and agroecological approaches that incorporate a diversity of crops and farming practices can enhance soil health, promote biodiversity, and reduce the reliance on chemical inputs like fertilizers and pesticides [44].
- **Eco-friendly Processing**: The processing of mucilage and gum products often involves various chemical treatments and industrial processes that may have environmental consequences. For example, the extraction of guar gum may involve the use of chemicals such as sodium hydroxide, which can generate waste products and contribute to water pollution if not properly managed. Similarly, the processing of gum arabic can result in wastewater production and emissions from drying operations. To reduce these environmental impacts, companies in the mucilage and gum industry are increasingly adopting more eco-friendly and sustainable production methods, such as using biodegradable solvents and reducing energy consumption in drying and processing stages [45].
- **Waste Management**: The production of mucilages and gums generates organic waste, which, if not properly managed, can contribute to environmental pollution. However, there is potential for these waste materials to be utilized as bioproducts or as raw materials for other industries. For instance, guar meal, a by-product of guar gum extraction, can be used as animal feed or as a source of

protein for human consumption. Similarly, waste from the processing of okra mucilage can be composted to enhance soil quality or used in the production of bioplastics. As demand for natural gums and mucilages continues to grow, sustainable agricultural practices, eco-friendly processing methods, and responsible consumption are essential to ensure that these valuable resources can be used without compromising human health or the environment.

6.7.1 Global Policy and Regulatory Landscape

The global policy and regulatory landscape surrounding mucilages and gums, like many other naturally derived ingredients, is shaped by a combination of food safety regulations, environmental protection laws, and industry standards.

Different countries and regions have their own regulatory authorities and safety standards that oversee the use of mucilages and gums. These policies reflect the need to balance the benefits these substances provide, with ensuring consumer safety and minimizing any potential harm to public health and the environment.

6.7.2 Key Regulatory Authorities

– **Food and Drug Administration (FDA)**:
In the United States, the **FDA** is the primary regulatory body responsible for overseeing the safety of food additives, including mucilages and gums. The FDA classifies substances like gum arabic, guar gum, and xanthan gum as "Generally Recognized as Safe" (GRAS), meaning they are considered safe for human consumption when used in accordance with established guidelines. The GRAS status allows manufacturers to use these substances in food products without requiring pre-market approval, provided that the substances meet the FDA's safety standards.

– **European Food Safety Authority (EFSA)**:
In the European Union, the **European Food Safety Authority** (EFSA) is tasked with assessing food-related risks and providing scientific advice on food safety issues. EFSA plays a crucial role in determining the acceptable daily intake (ADI) for gums and mucilages, ensuring that their use in food products does not pose any significant health risks. While many gums like *guar gum, xanthan gum*, and *gellan gum* are widely accepted, EFSA continually monitors new research and provides updated recommendations based on scientific evidence.

– **World Health Organization (WHO)**:
The **WHO** provides international guidelines on the safety of food additives, including gums and mucilages, in collaboration with the **Food and Agriculture Organization (FAO)**. WHO's **Joint FAO/WHO Expert Committee on Food Additives (JECFA)** eval-

uates the safety of various food ingredients, offering global standards and recommendations for their use. JECFA has set permissible limits and safety thresholds for a variety of gums used in food products globally.

– Codex Alimentarius Commission:
The **Codex Alimentarius** (or "Food Code"), created by the WHO and FAO, sets international food safety standards, guidelines, and codes of practice for food products. Codex provides comprehensive guidance on the acceptable levels of mucilages and gums in food and beverages. It also establishes criteria for food labeling, ensuring transparency and consumer protection.

– Health Canada:
In Canada, the **Health Canada** regulatory authority, particularly the **Food Directorate** of the **Health Products and Food Branch**, ensures that food products containing gums and mucilages are safe for human consumption. Similar to the FDA, Health Canada evaluates the GRAS status of these substances and assesses any potential health risks associated with their consumption.

– Australian and New Zealand Food Standards (FSANZ):
Food Standards Australia New Zealand (FSANZ) is responsible for regulating the use of gums and mucilages in food products in both Australia and New Zealand. FSANZ establishes maximum allowable concentrations of food additives, including gums, and ensures that their usage aligns with international safety standards.

6.7.3 Food Safety and Health Regulations

6.7.3.1 Food Additive Safety
The primary concern regarding mucilage and gum use in food is their safety for human health. Regulatory agencies often conduct rigorous evaluations of the toxicological data to establish safe usage levels. Safety testing generally includes:
- **Acute Toxicity Testing**: Determining the immediate harmful effects of excessive consumption.
- **Chronic Toxicity**: Evaluating the effects of long-term exposure to these substances, including potential carcinogenic, mutagenic, or reproductive effects.
- **Allergy Testing**: Some gums, like guar and acacia, can trigger allergic reactions in certain individuals. Regulatory bodies set limits on the levels of such ingredients, and may require additional labeling to protect vulnerable groups.

6.7.3.2 Labeling Requirements
Many regulatory bodies have strict guidelines regarding the labeling of products containing mucilages and gums. In the European Union, for example, food products must list all ingredients, including mucilages and gums, by their specific chemical names

(e.g., E412 for guar gum). Similarly, in the United States, food manufacturers are required to list these ingredients in the product's ingredients list. In the case of allergens, warnings must be displayed to alert consumers to the presence of any gum or mucilage that could cause allergic reactions.

6.7.3.3 Environmental Impact and Sustainability

As concerns about sustainability and environmental impact grow, regulations governing the environmental footprint of mucilage and gum production are becoming increasingly important. Several factors influence the eco-safety of mucilages and gums:

– **Water Usage**: Cultivating plants like guar gum, okra, and acacia requires significant water resources. In arid regions, the excessive water usage for gum production can strain local ecosystems. Therefore, regulatory bodies in water-scarce regions are implementing policies to limit water usage for nonessential industries.
– **Pesticides and Chemicals**: Many gum-producing crops, especially those grown in monocultures, require pesticides and chemical fertilizers. Countries with strong environmental protection laws, such as the European Union, impose strict regulations on pesticide residues in food products, ensuring that gums derived from such crops are free from harmful chemicals.

6.7.3.4 Traceability and Transparency

Regulatory frameworks are increasingly focused on improving traceability and transparency in the supply chain of natural gums and mucilages. This includes tracking the source of raw materials, assessing the ecological impact of cultivation practices, and ensuring ethical sourcing of these substances. Global certification standards such as **Fair Trade** and **Rainforest Alliance** are becoming more important in the global supply chain of gums, especially for those derived from tropical or sensitive ecosystems.

6.8 Challenges in Global Regulation

While regulatory frameworks are in place in many countries, there are several challenges that the global regulatory landscape faces in terms of mucilage and gum safety:

Variations in Standards: Different countries and regions have varying standards for the safety of gums and mucilages. What is considered safe in one country may be restricted or banned in another. This can create difficulties for international companies trying to market products globally. For example, guar gum is widely accepted in the United States but faces stricter regulations in certain parts of Europe.

Emerging Risks: As new mucilage and gum sources are identified, regulatory bodies must conduct rigorous testing to determine their safety. Certain gums from lesser-known plants may not yet have been adequately assessed, leading to regulatory gaps. Additionally, as new scientific data emerges, regulatory standards must be updated, which can be a slow and reactive process.

Ethical Sourcing: Ensuring that mucilage and gum production does not harm ecosystems or violate human rights remains a key concern. As the demand for natural gums increases, there is the risk of overharvesting and deforestation, particularly in the case of acacia gum and guar gum. Regulatory frameworks are increasingly focused on sustainability, but enforcement in developing countries remains a challenge.

References

[1]   Gowthaman, N. S. K., Lim, H. N., Sreeraj, T. R., Amalraj, A., & Gopi, S. (2021). Advantages of biopolymers over synthetic polymers: social, economic, and environmental aspects. In *Biopolymers and their industrial applications* (pp. 351–372). Elsevier.

[2]   Hahn, S., & Hennecke, D. (2023). What can we learn from biodegradation of natural polymers for regulation?. *Environmental Sciences Europe, 35*(1), 50.

[3]   Satchanska, G., Davidova, S., & Petrov, P. D. (2024). Natural and synthetic polymers for biomedical and environmental applications. *Polymers, 16*(8), 1159.

[4]   Jung, K., Corrigan, N., Wong, E. H., & Boyer, C. (2022). Bioactive synthetic polymers. Advanced Materials, 34(2), 2105063.

[5]   Moore, C. J. (2008). Synthetic polymers in the marine environment: a rapidly increasing, long-term threat. *Environmental research, 108*(2), 131–139.

[6]   Haider, T. P., Völker, C., Kramm, J., Landfester, K., & Wurm, F. R. (2019). Plastics of the future? The impact of biodegradable polymers on the environment and on society. *Angewandte Chemie International Edition, 58*(1), 50–62.

[7]   Swift, G. (1993). Directions for environmentally biodegradable polymer research. *Accounts of chemical research, 26*(3), 105–110.

[8]   Ghanbarzadeh, B., & Almasi, H. (2013). Biodegradable polymers. *Biodegradation-life of science,* 141–185.

[9]   Bhatia, S. (2016). Natural polymers vs synthetic polymer. In *Natural polymer drug delivery systems: nanoparticles, plants, and algae* (pp. 95–118). Cham: Springer International Publishing.

[10] Daei, S., Mohtarami, F., & Pirsa, S. (2022). A biodegradable film based on carrageenan gum/ Plantago psyllium mucilage/red beet extract: physicochemical properties, biodegradability and water absorption kinetic. *Polymer Bulletin, 79*(12), 11317–11338.

[11] Shiam, M. A. H., Islam, M. S., Ahmad, I., & Haque, S. S. (2025). A review of plant-derived gums and mucilages: Structural chemistry, film forming properties and application. *Journal of Plastic Film & Sheeting, 41*(2), 195–237.

[12] Beikzadeh, S., Khezerlou, A., Jafari, S. M., Pilevar, Z., & Mortazavian, A. M. (2020). Seed mucilages as the functional ingredients for biodegradable films and edible coatings in the food industry. *Advances in colloid and interface science, 280*, 102164.

[13] Malabadi, R. B., Kolkar, K. P., & Chalannavar, R. K. (2021). Natural plant gum exudates and mucilage: pharmaceutical updates. *Int J Innov Sci Res Rev, 3*(10), 1897–1912.

[14] Tosif, M. M., Najda, A., Bains, A., Zawiślak, G., Maj, G., & Chawla, P. (2021). Starch–mucilage composite films: An inclusive on physicochemical and biological perspective. *Polymers, 13*(16), 2588.

[15] Matei, E., Predescu, A. M., Râpă, M., Țurcanu, A. A., Mateș, I., Constantin, N., & Predescu, C. (2022). Natural polymers and their nanocomposites used for environmental applications. *Nanomaterials, 12*(10), 1707.

[16] Toma, D. I., Manaila-Maximean, D., Fierascu, I., Baroi, A. M., Matei, R. I., Fistos, T., . . . & Fierascu, R. C. (2024). Applications of natural polymers in the grapevine industry: Plant protection and value-added utilization of waste. *Polymers, 17*(1), 18.

[17] Mansor, M. R., Salit, M. S., Zainudin, E. S., Aziz, N. A., & Ariff, H. (2015). Life cycle assessment of natural fiber polymer composites. In *Agricultural biomass based potential materials* (pp. 121–141). Cham: Springer International Publishing.

[18] Groh, K. J., Arp, H. P. H., MacLeod, M., & Wang, Z. (2023). Assessing and managing environmental hazards of polymers: historical development, science advances and policy options. *Environmental Science: Processes & Impacts, 25*(1), 10–25.

[19] Ramesh, P., & Vinodh, S. (2020). State of art review on Life Cycle Assessment of polymers. *International Journal of Sustainable Engineering, 13*(6), 411–422.

[20] La Rosa, A. D., Recca, G., Summerscales, J., Latteri, A., Cozzo, G., & Cicala, G. (2014). Bio-based versus traditional polymer composites. A life cycle assessment perspective. *Journal of cleaner production, 74*, 135–144.

[21] Patel, M., Bastioli, C., Marini, L., & Würdinger, E. (2002). Environmental assessment of bio-based polymers and natural fibres. *Netherlands: Utrecht University*.

[22] Diana, Z., Vegh, T., Karasik, R., Bering, J., Caldas, J. D. L., Pickle, A., . . . & Virdin, J. (2022). The evolving global plastics policy landscape: An inventory and effectiveness review. *Environmental Science & Policy, 134*, 34–45.

[23] Chauhan, K., Kaur, R., & Chauhan, I. (2024). Sustainable bioplastic: a comprehensive review on sources, methods, advantages, and applications of bioplastics. *Polymer-Plastics Technology and Materials, 63*(8), 913–938.

[24] Angelopoulou, P., Giaouris, E., & Gardikis, K. (2022). *Applications and prospects of nanotechnology in food and cosmetics preservation. Nanomaterials. 2022; 12*: 1196.

[25] Toldy, A. (2023). Challenges and opportunities of polymer recycling in the changing landscape of European legislation. *Express Polymer Letters, 17*(11), 1081–1081.

[26] Khoo, P. S., Ilyas, R. A., Uda, M. N. A., Hassan, S. A., Nordin, A. H., Norfarhana, A. S., . . . & Rafiqah, S. A. (2023). Starch-based polymer materials as advanced adsorbents for sustainable water treatment: current status, challenges, and future perspectives. *Polymers, 15*(14), 3114.

[27] Jahan, I., & Zhang, L. (2022). Natural polymer-based electrospun nanofibrous membranes for wastewater treatment: A review. *Journal of Polymers and the Environment, 30*(5), 1709–1729.

[28] Taib, N. A. A. B., Rahman, M. R., Huda, D., Kuok, K. K., Hamdan, S., Bakri, M. K. B., . . . & Khan, A. (2023). A review on poly lactic acid (PLA) as a biodegradable polymer. *Polymer Bulletin, 80*(2), 1179–1213.

[29] Lee, S. Y., & Choi, J. I. (1999). Production and degradation of polyhydroxyalkanoates in waste environment. *Waste Management, 19*(2), 133–139.

[30] Ali, G., Sharma, M., Salama, E. S., Ling, Z., & Li, X. (2024). Applications of chitin and chitosan as natural biopolymer: potential sources, pretreatments, and degradation pathways. *Biomass Conversion and Biorefinery, 14*(4), 4567–4581.

[31] Kothale, D., Verma, U., Dewangan, N., Jana, P., Jain, A., & Jain, D. (2020). Alginate as promising natural polymer for pharmaceutical, food, and biomedical applications. *Current drug delivery, 17*(9), 755–775.

[32] Hamilton, J. D., Reinert, K. H., Hagan, J. V., & Lord, W. V. (1995). Polymers as solid waste in municipal landfills. *Journal of the Air & Waste Management Association, 45*(4), 247–251.

[33] Ayilara, M. S., Olanrewaju, O. S., Babalola, O. O., & Odeyemi, O. (2020). Waste management through composting: Challenges and potentials. *Sustainability*, *12*(11), 4456.

[34] Jahan, I., & Zhang, L. (2022). Natural polymer-based electrospun nanofibrous membranes for wastewater treatment: A review. *Journal of Polymers and the Environment*, *30*(5), 1709–1729.

[35] Fierascu, R. C., Fierascu, I., Matei, R. I., & Manaila-Maximean, D. (2023). Natural and natural-based polymers: Recent developments in management of emerging pollutants. *Polymers*, *15*(9), 2063.

[36] Korte, I., Kreyenschmidt, J., Wensing, J., Bröring, S., Frase, J. N., Pude, R., . . . & Schulze, M. (2021). Can sustainable packaging help to reduce food waste? A status quo focusing plant-derived polymers and additives. *Applied Sciences*, *11*(11), 5307.

[37] Angelopoulou, P., Giaouris, E., & Gardikis, K. (2022). *Applications and prospects of nanotechnology in food and cosmetics preservation. Nanomaterials. 2022; 12*: 1196.

[38] Nazarzadeh Zare, E., Mudhoo, A., Ali Khan, M., Otero, M., Bundhoo, Z. M. A., Patel, M., . . . & Sillanpää, M. (2021). Smart adsorbents for aquatic environmental remediation. *Small*, *17*(34), 2007840.

[39] Cocîrlea, M. D., Simionescu, N., Petrovici, A. R., Silion, M., Biondi, B., Lastella, L., & Oancea, S. (2024). In vitro screening of ecotoxic and cytotoxic activities of Ailanthus altissima leaf extract against target and non-target plant and animal cells. *International Journal of Molecular Sciences*, *25*(11), 5653.

[40] Chauhan, K., Kaur, R., & Chauhan, I. (2024). Sustainable bioplastic: a comprehensive review on sources, methods, advantages, and applications of bioplastics. *Polymer-Plastics Technology and Materials*, *63*(8), 913–938.

[41] Tawiah, B., Badoe, W., & Fu, S. (2016). Advances in the development of antimicrobial agents for textiles: The quest for natural products. Review. *Fibres & Textiles in Eastern Europe*, (3 (117), 136–149.

[42] Ma, S. Q., Cao, H. H., Wang, J. X., & Lin, S. (2014). Study on Eco-Safety Early-Warning and Assessment Index System of Hainan Province. In *Ecosystem Assessment and Fuzzy Systems Management* (pp. 33–43). Cham: Springer International Publishing.

[43] Abd-Elsalam, K. A., Alghuthaymi, M. A., & Abdel-Momen, S. M. (2023). Eco-safety and future trends.

[44] Jat, R. A., Wani, S. P., Sahrawat, K. L., & Singh, P. (2012). Eco-Safety and Agricultural Sustainability through Organic Agriculture.

[45] Abd-Elsalam, K. A., Alghuthaymi, M. A., & Abdel-Momen, S. M. (2023). Eco-safety and future trends.

7 Pharmaceutical Applications

7.1 Introduction

Mucilage and gums, naturally occurring polysaccharides, play a vital role in pharmaceutical formulations due to their versatile functional and therapeutic properties [1]. They are widely used as excipients in the development of tablets, suspensions, emulsions, and topical preparations [2]. As binding agents, they enhance tablet cohesion, ensuring proper compression and mechanical strength [1, 2]. Their swelling and gel-forming abilities make them ideal disintegrants, facilitating rapid tablet breakdown and improved drug release [3]. In suspensions and emulsions, mucilage and gums act as stabilizers and thickening agents, maintaining uniform drug distribution and enhancing product consistency [1, 3, 4]. Their bioadhesive properties are especially beneficial in controlled drug delivery systems, where prolonged residence time improves drug absorption and efficacy [1, 2]. Additionally, they serve as natural film-formers in coating applications, offering protection to active ingredients from environmental degradation [1]. Their biocompatibility, biodegradability, and non-toxicity make them suitable alternatives to synthetic polymers [3]. Furthermore, mucilage and gums have shown promise in wound healing and anti-inflammatory applications due to their soothing and protective nature [5]. Owing to their affordability, safety, and multifunctionality, these natural materials are increasingly favored in the formulation of modern and herbal pharmaceutical products, contributing significantly to drug delivery innovations and sustainable healthcare solutions [1].

7.2 Use of Mucilage and Gums in Drug Delivery Systems

Mucilages are a type of polysaccharide that absorbs water and swells, forming a gel-like substance. They are commonly derived from plant seeds, roots, and other plant parts. Examples of mucilages include okra mucilage, flaxseed mucilage, and guar gum. Gums, on the other hand, are complex polysaccharides or glycoproteins that are soluble or swell in water and form viscous solutions or gels [6].

7.3 Roles of Mucilage and Gums in Drug Delivery Systems

7.3.1 Controlled and Sustained Drug Release

One of the key applications of mucilages and gums in pharmaceutical formulations is their ability to control the release of active pharmaceutical ingredients (APIs). The release rate of a drug is crucial in determining its therapeutic effectiveness and safety.

https://doi.org/10.1515/9783111673509-007

Drugs that are rapidly released may lead to toxic levels in the bloodstream, while drugs with slow release may not achieve their intended therapeutic effect. Mucilages and gums can be utilized to create controlled-release (CR) and sustained-release dosage forms by modifying the release kinetics [7]:

- **Hydrogel-Based Drug Delivery**: Mucilages and gums, due to their ability to absorb water and form gels, are used in the preparation of hydrogel matrices. These gels can control the release of drugs by slowing their diffusion through the gel network. For example, xanthan gum is commonly used in hydrogel-based drug delivery systems because of its high viscosity and ability to form stable gels. When used in matrix systems, xanthan gum can control the release of hydrophobic drugs, such as ibuprofen, over extended periods [8].
- **Gum-Based CR Tablets**: Gums like guar gum, tragacanth, and acacia gum are used in CR tablet formulations. These gums form a gel when in contact with water, creating a barrier around the API. This barrier slows the release of the drug over time. For instance, guar gum is often used in the formulation of sustained-release tablets of diclofenac sodium. The gum's ability to swell and retain water reduces the drug's rate of dissolution, leading to prolonged drug release [9].
- **Polymeric Microparticles and Nanoparticles**: Mucilages and gums are also utilized in the formulation of microparticles and nanoparticles for drug delivery. These microparticles or nanoparticles, often composed of xanthan gum, guar gum, or sodium alginate, can encapsulate both hydrophilic and hydrophobic drugs, controlling their release rate by modifying the size, surface charge, and permeability of the particles. For example, guar gum nanoparticles have been successfully used to encapsulate doxorubicin, achieving a sustained-release profile that reduces systemic side effects [10].

7.3.2 Mucoadhesive Drug Delivery

Mucoadhesive drug delivery systems are designed to adhere to the mucosal tissues in the body (e.g., the gastrointestinal tract, nasal passages, or vaginal mucosa) to prolong the residence time of the drug at the site of action. This results in improved drug absorption and localized therapeutic effects [11]:

- **Mucoadhesive Tablets**: Mucilages and gums are often incorporated into mucoadhesive tablets, which are designed to adhere to mucosal surfaces. These mucoadhesive systems provide prolonged contact with the mucosal lining, leading to enhanced bioavailability of the drug. Acacia gum and guar gum have been extensively studied for their mucoadhesive properties. These gums form a gel-like layer when hydrated, allowing them to adhere to mucosal surfaces, which prolongs the release and absorption of drugs [12].

- **Nasal Drug Delivery Systems**: Mucilages and gums are also used in nasal drug delivery systems, where they serve as mucoadhesive agents that help drugs adhere to the nasal mucosa. This is particularly useful for biologic drugs (e.g., insulin) that require rapid absorption. Xanthan gum and tragacanth gum are often incorporated into nasal sprays or gel formulations to enhance the contact time and absorption of the active ingredient in the nasal cavity [13].
- **Vaginal Drug Delivery**: In vaginal drug delivery, mucilages and gums can be used to design mucoadhesive pessaries or suppositories. These formulations are designed to release drugs such as antifungal agents or hormones at a controlled rate. The mucoadhesive properties of gums like guar gum and acacia gum allow for extended residence time in the vaginal mucosa, improving the local efficacy of the drug [14].

7.3.3 Stabilization of Liquid Dosage Forms

Gums and mucilages are frequently used in liquid pharmaceutical formulations to stabilize suspensions, emulsions, and solutions. These polysaccharides can increase the **viscosity** of liquid formulations, preventing the settling of suspended particles and maintaining the uniformity of the drug product [15]:
- **Suspensions and Emulsions**: Mucilages such as guar gum and xanthan gum are often used as suspending agents in liquid dosage forms. Their ability to increase viscosity helps maintain the uniform dispersion of active ingredients in suspension. For example, paracetamol suspensions are stabilized using guar gum, which prevents the drug particles from settling and ensures consistent dosing [16].
- **Emulsion Stability**: Gums such as **acacia gum** and **tragacanth** are commonly used in the preparation of **emulsions** (oil-in-water or water-in-oil systems). These gums act as **emulsifying agents**, helping to stabilize the dispersed phases and prevent phase separation. This is particularly important in the formulation of emulsified systems used in **topical** or **oral** drug delivery [17].

7.3.4 Targeted Drug Delivery

Gums and mucilages can also be used in targeted drug delivery systems, which aim to deliver the drug specifically to a desired location in the body, thus minimizing side effects and maximizing therapeutic outcomes. The use of natural gums in such systems is an area of active research, particularly in colonic drug delivery and site-specific drug release [18]:
- **Colon-Specific Drug Delivery**: One of the most promising applications of mucilage and gums is in colonic drug delivery. The ability of certain gums to degrade in the colon due to the presence of specific enzymes makes them ideal for colon-

targeted drug delivery. For example, guar gum and xanthan gum are used in colon-specific drug formulations, where they remain intact in the stomach and small intestine but are metabolized by colonic bacteria, releasing the drug in the colon. This system is beneficial for drugs used in the treatment of colonic diseases like ulcerative colitis or colon cancer [19].

– **pH-Dependent Release**: Certain gums, such as pectin and gellan gum, have pH-sensitive properties, making them ideal candidates for use in pH-dependent drug release systems. These gums remain stable in acidic environments (like the stomach) but begin to degrade in neutral to alkaline conditions (such as in the small intestine). This allows for the CR of drugs at specific sites in the gastrointestinal tract [20].

7.4 Role in Controlled and Targeted Release

CR drug delivery systems are designed to release the API in a manner that maintains its therapeutic concentration over an extended period, reducing the frequency of dosing and improving patient compliance. CR systems aim to release the drug in a predictable and sustained manner, typically by means of a matrix system, reservoir system, or osmotic system. By using biodegradable polymers like mucilages and gums, CR formulations can be engineered to ensure that the drug is available to the body in a controlled manner over an extended period, minimizing the risk of toxicity and optimizing therapeutic effectiveness [21].

Mechanisms of Controlled Drug Release Using Mucilages and Gums
The mechanisms by which mucilages and gums control the release of drugs can vary based on their chemical structure, water-solubility, and the characteristics of the drug being delivered.

7.4.1 Gel Formation and Swelling Behavior

When exposed to aqueous environments, mucilages and gums absorb water and undergo significant swelling, which causes the formation of a gel-like network. This property is fundamental to their role in CR. As the gum or mucilage swells, the drug encapsulated within it is gradually released through the gel network via diffusion. The swelling behavior is influenced by the degree of cross-linking, hydrophilicity, and gelation conditions (pH and temperature).

For example, guar gum, when incorporated in a CR formulation, swells upon contact with water, creating a gel matrix that encapsulates the drug. The drug then diffuses through the gel at a controlled rate, depending on the swelling properties of the gum and the viscosity of the gel. The release rate can be modulated by adjusting the concentration of the gum or pH-sensitive modifications to the polymer [22].

7.4.2 Matrix Systems: Gel Entrapment

The most common approach in CR using mucilages and gums is the matrix system, where the drug is incorporated into the polymer matrix. The drug is slowly released as it diffuses through the gel network. The release can be zero-order (constant release over time) or first-order (dependent on the amount of drug left in the system).

Xanthan gum and guar gum are frequently used to form matrix systems for CR formulations. These gums, when hydrated, form highly viscous solutions and gels that allow drugs to be entrapped and released at a controlled rate. For instance, ibuprofen encapsulated in a matrix containing guar gum can be released over several hours, ensuring prolonged therapeutic effects with fewer doses [23].

7.4.3 Erosion of the Matrix

Another mechanism for controlled drug release is the erosion of the polymer matrix. As the gum or mucilage interacts with water, it gradually erodes, allowing the drug to be released over time. This mechanism is especially common in hydrophilic polymers, which are capable of absorbing water and slowly degrading [24].

Acacia gum and tragacanth gum are known for their ability to form matrices that erode over time. In these systems, the polymer dissolves or erodes gradually, allowing the drug to be released at a controlled rate. For example, a tragacanth gum-based matrix could be used for the CR of an antibiotic like ampicillin, ensuring the drug is available over an extended period, even if the patient takes it only once or twice a day [23].

7.4.4 Diffusion-Controlled Systems

In diffusion-controlled systems, the drug release is governed by the diffusion of the drug through the gel or polymer matrix. The rate of drug release depends on the size and composition of the matrix, as well as the solubility of the drug.

Mucilages and gums that form highly cross-linked, hydrophilic networks, such as xanthan gum and pectin, are particularly effective in this type of system. For instance, a xanthan gum-based tablet may release a hydrophilic drug like acetaminophen by diffusion through the gel structure, allowing the drug to be released in a controlled manner over several hours [25].

Targeted drug delivery refers to the approach of delivering drugs directly to specific cells, tissues, or organs, improving the therapeutic efficacy of the drug and minimizing side effects. Targeted drug delivery systems are designed to release the drug only at the site of action, rather than throughout the body, to avoid systemic side effects and improve the bioavailability of poorly absorbed drugs [26].

Mucilages and gums have the potential to enhance targeted delivery due to their biodegradability, ability to be modified for site-specific release, and capacity for mucoadhesion (the ability to adhere to mucosal surfaces). The targeting mechanisms can be passive (based on the physicochemical properties of the polymer) or active (involving the use of ligands or antibodies to recognize and bind to specific receptors on the target cells).

7.4.5 Colonic Drug Delivery

One of the major applications of mucilages and gums in targeted delivery is in colonic drug delivery systems. The colon is a challenging site for drug absorption, but it is particularly suitable for drugs that are intended for local treatment (such as anti-inflammatory agents or chemotherapeutic drugs) or for drugs that undergo extensive first-pass metabolism in the liver.

Polymers like guar gum, xanthan gum, and pectin are highly beneficial in colon-specific drug delivery. These gums are biodegradable and can resist digestion in the stomach and small intestine. However, they are susceptible to microbial degradation in the colon, where certain bacteria break down the polysaccharides, releasing the drug at the target site [27].

For example, guar gum and xanthan gum can be used in tablets or capsules designed to pass through the stomach and small intestine intact, only to degrade in the colon due to the action of colonic bacteria. This property is particularly useful for colon-targeted formulations of anti-inflammatory drugs used in the treatment of inflammatory bowel disease (IBD) or ulcerative colitis.

7.4.6 Nasal and Buccal Drug Delivery

Mucilages and gums are also used in mucoadhesive drug delivery systems for the nasal, buccal, and vaginal routes. In these formulations, the gum or mucilage adheres to the mucosal lining, prolonging the residence time of the drug at the absorption site. This increases the bioavailability of drugs, especially those that undergo significant first-pass metabolism when taken orally.

Xanthan gum and tragacanth gum are frequently used in nasal sprays, buccal tablets, and other mucoadhesive systems for local drug delivery. For instance, insulin can be delivered via a buccal gel containing xanthan gum, allowing for improved absorption through the mucosal lining of the mouth without undergoing digestion in the gastrointestinal tract [28].

7.4.7 Surface Modifications for Active Targeting

Mucilages and gums can also be modified to actively target specific cells or tissues. For example, xanthan gum can be functionalized with ligands or antibodies that bind specifically to receptors expressed on cancer cells or inflamed tissues. These modified gums can then serve as carriers for chemotherapeutic drugs or anti-inflammatory agents, ensuring that the drug is released at the target site, improving therapeutic efficacy and reducing systemic toxicity [29].

7.5 Applications in Wound Healing, Tablet Formulation, and Biomedicine

Their multifunctional properties such as gelling, emulsifying, thickening, and binding make them ideal candidates for a broad spectrum of applications in the pharmaceutical and biomedical fields [1, 2].

7.5.1 Mucilage and Gums in Wound Healing

Wound healing is a complex biological process that involves the restoration of tissue integrity after injury. This process is influenced by various factors, including the type and extent of the injury, the presence of infection, and the physiological conditions of the individual. One of the most significant advancements in wound care has been the use of natural biopolymers such as mucilages and gums to enhance healing, reduce infection, and speed up the recovery process. The role of mucilages and gums in wound healing can be understood through their ability to form protective barriers, promote tissue regeneration, and deliver active agents to the wound site [30].

7.5.2 Barrier Formation and Protection

Mucilages and gums, owing to their hydrophilic nature, are often employed in the development of wound dressings that provide a moist environment around the wound. This is crucial for optimal healing, as moist wound healing has been shown to accelerate recovery and reduce scarring. Hydrocolloid dressings, which often contain guar gum, xanthan gum, or pectin, form a gel-like layer over the wound that not only protects the area from external contaminants but also maintains the necessary moisture levels for faster tissue regeneration [31].

For instance, guar gum-based dressings have been shown to provide a protective layer that adheres to the wound bed while allowing for the exchange of gases and moisture. These dressings also create an environment conducive to cell proliferation and tissue regeneration. Guar gum has also been studied for its ability to accelerate collagen deposition, a critical factor in wound healing. The collagen matrix provides structural support to the healing tissue, enhancing the reepithelialization process [32].

Similarly, xanthan gum, with its high viscosity, is often used in formulations designed to be applied as gels or ointments for wound healing. Its ability to absorb exudate from the wound and maintain a moist environment makes it an excellent choice for chronic wounds, such as diabetic ulcers or pressure sores. Furthermore, xanthan gum also has the ability to form a stable gel structure that resists shearing forces, ensuring that the gel remains in place over the wound for extended periods.

7.6 Antibacterial and Antimicrobial Properties

Wounds, especially those caused by burns or surgical procedures, are susceptible to infection. One of the significant benefits of mucilages and gums in wound healing applications is their antimicrobial properties, which help prevent infection. Certain gums and mucilages, such as guar gum and tragacanth gum, have been studied for their ability to act as antibacterial agents. These natural polymers can be used to deliver antimicrobial agents, such as silver nanoparticles or antibiotics, to the wound site, reducing the risk of infection and promoting faster healing [33].

For example, silver-loaded guar gum hydrogels have been developed for the treatment of infected wounds. The hydrophilic nature of guar gum allows it to incorporate silver nanoparticles, which exhibit significant antimicrobial activity against a wide range of pathogens, including *Staphylococcus aureus* and *Escherichia coli*. These formulations are particularly effective for treating burn wounds or surgical site infections, where the risk of infection is high [34].

Moreover, xanthan gum has been investigated for its ability to form biofilms that can support the local release of antimicrobial agents. This approach can be especially beneficial for the treatment of biofilm-associated infections, such as those seen in chronic wounds, where conventional antibiotic treatments often fail due to bacterial resistance.

7.7 Promotion of Tissue Regeneration

Mucilages and gums play a vital role in the regeneration of skin and other tissues by providing a scaffold for cell attachment and migration. Their biocompatibility ensures that they can be used in direct contact with living tissues without causing adverse reactions. In particular, pectin, guar gum, and xanthan gum have been found to support the proliferation of skin cells (keratinocytes) and fibroblasts, which are essential for wound closure and tissue repair [35].

For instance, pectin-based hydrogels have been shown to improve the migration and proliferation of fibroblasts in vitro, promoting the formation of new tissue at the wound site. The gelling property of pectin provides a three-dimensional matrix that mimics the extracellular matrix, offering a platform for cellular growth and collagen deposition. This is particularly advantageous for the regeneration of dermal tissues in deep wounds or burns [36].

Guar gum has also been studied for its role in stimulating angiogenesis, the process by which new blood vessels form, which is critical for wound healing. By promoting blood flow to the wound site, guar gum-based formulations may enhance nutrient delivery and accelerate tissue repair, particularly in wounds that are slow to heal, such as diabetic ulcers.

7.8 Mucilage and Gums in Tablet Formulation

Tablet formulations are one of the most common dosage forms used in pharmaceutical practice, and mucilages and gums have found extensive use as binders, disintegrants, and release modifiers in tablet manufacturing. Their ability to improve tablet properties, such as hardness, dissolution, and bioavailability, makes them invaluable in the development of oral dosage forms [37].

7.8.1 Binders in Tablet Formulation

Mucilages and gums are widely employed as binders in the preparation of tablets, ensuring that the powder ingredients stick together during the compression process. The viscosity of gums such as guar gum, acacia gum, and tragacanth gum allows for the formation of a cohesive mass when mixed with the APIs and excipients [38].

For instance, guar gum is commonly used in the formulation of tablet binders, especially for tablet coatings and matrix tablets. Guar gum's ability to form strong gels at low concentrations ensures that the tablet retains its integrity during manufacturing, while also aiding in the CR of the drug over time. The polymer also acts as a moisture-retaining agent, which helps preserve the physical stability of the tablet during storage.

Similarly, xanthan gum has been used in the preparation of sustained-release tablets, where it serves as a binder while controlling the rate of drug release. Xanthan gum's ability to form viscous solutions in water allows for the formation of hydrogel matrices, which can regulate drug release by controlling the diffusion of the API.

7.8.2 Disintegrants in Tablet Formulation

In addition to their use as binders, mucilages and gums are also used as disintegrants in tablet formulations. Disintegrants are substances that promote the breakup of the tablet when it comes into contact with fluids in the gastrointestinal tract, ensuring rapid release of the drug. Mucilages such as guar gum, xanthan gum, and pectin are known to swell upon hydration, leading to the rupture and disintegration of the tablet [39].

For example, guar gum and pectin are used in formulations designed for fast-dissolving tablets, where they facilitate the rapid disintegration and release of the drug in the stomach. The swelling property of these gums helps break the tablet apart, allowing for quick release of the active ingredient.

7.8.3 Release Modifiers in Tablet Formulation

Mucilages and gums also play an important role as release modifiers in the preparation of sustained-release and CR tablet formulations. By using gums like guar gum, xanthan gum, and acacia gum, pharmaceutical scientists can design tablets that release the drug over an extended period, reducing the need for frequent dosing [40].

For example, in sustained-release tablets of theophylline, guar gum is used to form a matrix that controls the rate at which the drug is released into the bloodstream. The viscosity and swelling properties of guar gum enable it to regulate the diffusion of the drug, providing a steady release over several hours.

7.9 Mucilage and Gums in Biomedicine

In the broader field of **biomedicine**, mucilages and gums have found various applications, from drug delivery systems to tissue engineering. Their **biocompatibility** and **biodegradability** make them ideal candidates for use in applications that require direct interaction with living tissues [41].

7.9.1 Drug Delivery Systems

In biomedical drug delivery, mucilages and gums are used to create formulations that can deliver drugs to specific sites in the body, ensuring better therapeutic outcomes. For instance, xanthan gum and guar gum are used in mucoadhesive drug delivery systems, which target the gastrointestinal or nasal mucosa to provide localized drug delivery [42].

For example, guar gum is often used in oral drug delivery systems to enhance the bioavailability of poorly absorbed drugs. By forming a gel at the site of administration, it increases the retention time of the drug at the absorption site, allowing for improved intestinal absorption.

7.9.2 Tissue Engineering and Regenerative Medicine

Mucilages and gums also have applications in tissue engineering and regenerative medicine, where they serve as scaffolds for cell growth and tissue regeneration. Polymers such as pectin and xanthan gum have been investigated for their ability to provide structural support for cells in the process of tissue regeneration [43].

Pectin-based hydrogels have been used as scaffolds for cartilage and bone tissue engineering. These hydrogels provide a suitable environment for cell attachment, proliferation, and differentiation, making them ideal for use in regenerative medicine [44].

References

[1] Sangwan, Y. S., Sngwan, S., Jalwal, P., Murti, K., & Kaushik, M. (2011). Mucilages and their pharmaceutical applications: an overview. *Pharmacologyonline, 2,* 1265–1271.
[2] Prajapati, V., Desai, S., Gandhi, S., & Roy, S. (2022). Pharmaceutical applications of various natural gums and mucilages. In *Gums, resins and latexes of plant origin: Chemistry, biological activities and uses* (pp. 25–57). Cham: Springer International Publishing.
[3] Bahadur, S., Sahu, U. K., Sahu, D., Sahu, G., & Roy, A. (2017). Review on natural gums and mucilage and their application as excipient. *Journal of applied pharmaceutical research, 5*(4), 13–21.
[4] Malabadi, R. B., Kolkar, K. P., & Chalannavar, R. K. (2021). Natural plant gum exudates and mucilage: pharmaceutical updates. *Int J Innov Sci Res Rev, 3*(10), 1897–1912.
[5] Jani, G. K., Shah, D. P., Prajapati, V. D., & Jain, V. C. (2009). Gums and mucilages: versatile excipients for pharmaceutical formulations. *Asian J Pharm Sci, 4*(5), 309–323.
[6] Bhosale, R. R., Osmani, R. A. M., & Moin, A. (2014). Natural gums and mucilages: a review on multifaceted excipients in pharmaceutical science and research. *International Journal of Pharmacognosy and Phytochemical Research, 15*(6), 4.
[7] Avachat, A. M., Dash, R. R., & Shrotriya, S. N. (2011). Recent investigations of plant based natural gums, mucilages and resins in novel drug delivery systems. *Ind J Pharm Edu Res, 45*(1), 86–99.
[8] Amiri, M. S., Mohammadzadeh, V., Yazdi, M. E. T., Barani, M., Rahdar, A., & Kyzas, G. Z. (2021). Plant-based gums and mucilages applications in pharmacology and nanomedicine: a review. *Molecules, 26*(6), 1770.
[9] Kumar, S., & Gupta, S. K. (2012). Natural polymers, gums and mucilages as excipients in drug delivery. *Polim. Med, 42*(3–4), 191–197.
[10] Shiam, M. A. H., Islam, M. S., Ahmad, I., & Haque, S. S. (2025). A review of plant-derived gums and mucilages: Structural chemistry, film forming properties and application. *Journal of Plastic Film & Sheeting, 41*(2), 195–237.

[11]   Malviya, R., Srivastava, P., & Kulkarni, G. T. (2011). Applications of mucilages in drug delivery-a review. *Advances in Biological Research*, *5*(1), 1–7.

[12]   Sabale, V., Patel, V., & Paranjape, A. (2014). Evaluation of Calendula mucilage as a mucoadhesive and controlled release component in buccal tablets. *Research in pharmaceutical sciences*, *9*(1), 39–48.

[13]   Basu, S., & Bandyopadhyay, A. K. (2010). Development and characterization of mucoadhesive in situ nasal gel of midazolam prepared with Ficus carica mucilage. *Aaps Pharmscitech*, *11*(3), 1223–1231.

[14]   Veiga, M. D., Ruiz-Caro, R., Martín-Illana, A., Notario-Pérez, F., & Cazorla-Luna, R. (2018). Polymer gels in vaginal drug delivery systems. In *Polymer Gels: Synthesis and Characterization* (pp. 197–246). Singapore: Springer Singapore.

[15]   Jani, G. K., Shah, D. P., Prajapati, V. D., & Jain, V. C. (2009). Gums and mucilages: versatile excipients for pharmaceutical formulations. *Asian J Pharm Sci*, *4*(5), 309–323.

[16]   Avlani, D., Agarwal, V. A. I. B. H. A. V., Khattry, V., Biswas, G. R., & Majee, S. B. (2019). Exploring properties of sweet basil seed mucilage in development of pharmaceutical suspensions and surfactant-free stable emulsions. *Int J Appl Pharm*, *11*(1), 124–129.

[17]   Koocheki, A., Taherian, A. R., Razavi, S. M., & Bostan, A. (2009). Response surface methodology for optimization of extraction yield, viscosity, hue and emulsion stability of mucilage extracted from Lepidium perfoliatum seeds. *Food Hydrocolloids*, *23*(8), 2369–2379.

[18]   Amiri, M. S., Mohammadzadeh, V., Yazdi, M. E. T., Barani, M., Rahdar, A., & Kyzas, G. Z. (2021). Plant-based gums and mucilages applications in pharmacology and nanomedicine: a review. *Molecules*, *26*(6), 1770.

[19]   Jain, A., Gupta, Y., & Jain, S. K. (2007). Perspectives of biodegradable natural polysaccharides for site-specific drug delivery to the colon. *J Pharm Pharm Sci*, *10*(1), 86–128.

[20]   Sarfraz, M., Tulain, U. R., Erum, A., Malik, N. S., Mahmood, A., Aslam, S., . . . & Tayyab, M. (2023). Cydonia oblonga-Seed-Mucilage-Based pH-Sensitive Graft Copolymer for Controlled Drug Delivery – In Vitro and In Vivo Evaluation. *Pharmaceutics*, *15*(10), 2445.

[21]   Tosif, M. M., Najda, A., Bains, A., Kaushik, R., Dhull, S. B., Chawla, P., & Walasek-Janusz, M. (2021). A comprehensive review on plant-derived mucilage: characterization, functional properties, applications, and its utilization for nanocarrier fabrication. *Polymers*, *13*(7), 1066.

[22]   Brax, M., Schaumann, G. E., & Diehl, D. (2019). Gel formation mechanism and gel properties controlled by Ca2+ in chia seed mucilage and model substances. *Journal of Plant Nutrition and Soil Science*, *182*(1), 92–103.

[23]   Nokhodchi, A., Nazemiyeh, H., Khodaparast, A., Sorkh-Shahan, T., Valizadeh, H., & Ford, J. L. (2008). An in vitro evaluation of fenugreek mucilage as a potential excipient for oral controlled-release matrix tablet. *Drug development and industrial pharmacy*, *34*(3), 323–329.

[24]   Fernandes, S. S., Da silva cardoso, P., Egea, M. B., Martínez, J. P. Q., Campos, M. R. S., & Otero, D. M. (2023). Chia mucilage carrier systems: A review of emulsion, encapsulation, and coating and film strategies. *Food Research International*, *172*, 113125.

[25]   Malviya, R., Tyagi, V., & Singh, D. (2020). Techniques of mucilage and gum modification and their effect on hydrophilicity and drug release. *Recent Patents on Drug Delivery & Formulation*, *14*(3), 214–222.

[26]   Tosif, M. M., Najda, A., Bains, A., Kaushik, R., Dhull, S. B., Chawla, P., & Walasek-Janusz, M. (2021). A comprehensive review on plant-derived mucilage: characterization, functional properties, applications, and its utilization for nanocarrier fabrication. *Polymers*, *13*(7), 1066.

[27]   Shahid, M., Munir, H., Akhter, N., Akram, N., Anjum, F., Iqbal, Y., & Afzal, M. (2021). Nanoparticles encapsulation of Phoenix dactylifera (date palm) mucilage for colonic drug delivery. *International Journal of Biological Macromolecules*, *191*, 861–871.

[28]   Basu, S., & Bandyopadhyay, A. K. (2010). Development and characterization of mucoadhesive in situ nasal gel of midazolam prepared with Ficus carica mucilage. *Aaps Pharmscitech*, *11*(3), 1223–1231.

[29] Liu, Y., Liu, Z., Zhu, X., Hu, X., Zhang, H., Guo, Q., . . . & Cui, S. W. (2021). Seed coat mucilages: Structural, functional/bioactive properties, and genetic information. *Comprehensive Reviews in Food Science and Food Safety, 20*(3), 2534–2559.

[30] Adikwu, M. U., & Enebeke, T. C. (2007). Evaluation of snail mucin dispersed in Brachystegia gum gel as a wound healing agent. *Animal research international, 4*(2), 685–697.

[31] Tee, Y. B., Tee, L. T., Daengprok, W., & Talib, R. A. (2017). Chemical, physical, and barrier properties of edible film from flaxseed mucilage. *BioResources, 12*(3), 6656–6664.

[32] Li, Y., Duan, Q., Yue, S., Alee, M., & Liu, H. (2024). Enhancing mechanical and water barrier properties of starch film using chia mucilage. *International Journal of Biological Macromolecules, 274*, 133288.

[33] Mohammadi, H., Kamkar, A., & Misaghi, A. (2018). Nanocomposite films based on CMC, okra mucilage and ZnO nanoparticles: Physico mechanical and antibacterial properties. *Carbohydrate Polymers, 181*, 351–357.

[34] Tantiwatcharothai, S., & Prachayawarakorn, J. (2019). Characterization of an antibacterial wound dressing from basil seed (Ocimum basilicum L.) mucilage-ZnO nanocomposite. *International journal of biological macromolecules, 135*, 133–140.

[35] Aghmiuni, A. I., Keshel, S. H., Sefat, F., & Khiyavi, A. A. (2020). Quince seed mucilage-based scaffold as a smart biological substrate to mimic mechanobiological behavior of skin and promote fibroblasts proliferation and h-ASCs differentiation into keratinocytes. *International journal of biological macromolecules, 142*, 668–679.

[36] Kumari, P., Ahina, K. M., Kannan, K., Sreekumar, S., Lakra, R., Sivagnanam, U. T., & Kiran, M. S. (2024). In vivo soft tissue regenerative potential of flax seed mucilage self-assembled collagen aerogels. *Biomedical Materials, 19*(2), 025023.

[37] Arshi, A. (2011). *Formulation and evaluation of tablets using natural gum as a binder* (Master's thesis, Rajiv Gandhi University of Health Sciences (India)).

[38] Ahuja, M., Kumar, A., Yadav, P., & Singh, K. (2013). Mimosa pudica seed mucilage: Isolation; characterization and evaluation as tablet disintegrant and binder. *International journal of biological macromolecules, 57*, 105–110.

[39] Tel, D., Prajapati, D. G., & Patel, N. M. (2007). Seed mucilage from Ocimum americanum linn. as disintegrant in tablets: Separation and evaluation. *Indian Journal of pharmaceutical sciences*.

[40] Srivastava, P., Malviya, R., Gupta, S., & Sharma, P. K. (2010). Evaluation of various natural Gums as release modifiers in tablet formulations. *Pharmacognosy Journal, 2*(13), 525–529.

[41] Amiri, M. S., Mohammadzadeh, V., Yazdi, M. E. T., Barani, M., Rahdar, A., & Kyzas, G. Z. (2021). Plant-based gums and mucilages applications in pharmacology and nanomedicine: a review. *Molecules, 26*(6), 1770.

[42] Krishna, L. N. V., Kulkarni, P. K., Dixit, M., Lavanya, D., & Raavi, P. K. (2011). Brief introduction of natural gums, mucilages and their applications in novel drug delivery systems-a review. *IJDFR, 2*(6), 54–71.

[43] Farshidfar, N., Iravani, S., & Varma, R. S. (2023). Alginate-based biomaterials in tissue engineering and regenerative medicine. *Marine Drugs, 21*(3), 189.

[44] Abualsoud, B. M., Alhomrani, M., Alamri, A. S., Alsanie, W. F., Baldaniya, L., Jyothi, S. R., . . . & A, D. (2025). Pectin-based composites: a promising approach for tissue engineering and wound healing. *International Journal of Polymeric Materials and Polymeric Biomaterials*, 1–28.

8 Food and Beverage Industry

8.1 Introduction

Mucilages and gums play a significant role in the food and beverage industry due to their exceptional functional and rheological properties [1]. These natural hydrocolloids act as stabilizers, thickeners, emulsifiers, and gelling agents, contributing to the texture, consistency, and shelf life of various food products. In beverages, gums like guar gum, xanthan gum, and mucilage from plant sources are used to maintain uniform dispersion of ingredients, prevent sedimentation, and enhance mouthfeel [2]. In processed foods, they provide viscosity and stability, improving the appearance and acceptability of sauces, dressings, dairy products, and baked goods. Mucilage also assists in fat replacement by mimicking the texture of fats, thus promoting the development of low-calorie and health-oriented food options [1, 3]. Their ability to retain moisture helps in prolonging freshness and reducing staleness in bakery items. Additionally, gums and mucilage function as dietary fibers, supporting digestive health and contributing to functional foods. Their natural origin, safety, and biodegradability make them suitable for clean-label formulations and health-conscious consumers [4]. As demand for plant-based, sustainable ingredients grows, mucilages and gums continue to gain importance as multifunctional additives, offering both technological benefits and nutritional value in a wide range of food and beverage applications.

8.2 Natural Thickeners and Stabilizers in Food Products

Natural thickeners and stabilizers are substances used in food processing to alter the texture, consistency, and stability of food products. Thickeners increase the viscosity of a product without significantly affecting its other properties, such as taste or nutritional profile. Stabilizers, on the other hand, help maintain the homogeneity of a mixture by preventing the separation of ingredients, such as oil and water in emulsions. These ingredients play a crucial role in enhancing the mouthfeel, appearance, and shelf life of food products. While synthetic thickeners and stabilizers have been widely used in food manufacturing, natural alternatives are gaining popularity due to their biocompatibility, sustainability, and alignment with consumer preferences for natural, clean-label ingredients [5].

Natural thickeners and stabilizers are primarily derived from various plant sources, marine organisms, and even microorganisms. Some of the most common natural thickeners and stabilizers include **gums** (guar gum, xanthan gum, locust bean gum, and gum arabic), pectin, starches (cornstarch, potato starch, and arrowroot), and seaweed-derived products such as agar-agar, carrageenan, and alginate. These biopolymers are utilized for their unique functional properties, which enable food manufac-

https://doi.org/10.1515/9783111673509-008

turers to enhance the sensory characteristics of food, improve texture, and maintain product stability during storage [6].

8.2.1 Types of Natural Thickeners and Stabilizers

8.2.1.1 Gums
Gums are among the most widely used natural thickeners and stabilizers in the food industry. These substances are primarily polysaccharides that exhibit excellent water-binding properties. When dissolved in water, gums can form gels or pastes that are effective at increasing the viscosity of liquids and providing a thick, smooth consistency [7]:

- **Guar Gum**: Derived from the seeds of the **guar plant** (*Cyamopsis tetragonoloba*), guar gum is a high-molecular-weight polysaccharide, known for its ability to significantly increase the viscosity of aqueous solutions even at low concentrations. Guar gum is widely used in **dairy products**, **beverages**, and **sauces**, where it helps improve texture and prevent phase separation. It is also used in **gluten-free baking** as a binder and moisture-retaining agent [8].
- **Xanthan Gum**: Produced by the fermentation of **glucose** or **sucrose** using the bacterium *Xanthomonas campestris*, xanthan gum is a versatile gum used in a wide range of food applications. It is particularly effective as an **emulsifier**, stabilizing emulsions such as salad dressings, sauces, and mayonnaise. It also serves as a **thickener** in beverages, soups, and gravies, and helps improve **mouthfeel** in low-fat food products by mimicking the texture provided by fats [9].
- **Locust Bean Gum**: Extracted from the seeds of the **carob tree** (*Ceratonia siliqua*), locust bean gum is a galactomannan that is commonly used in combination with other gums, such as **xanthan gum** and **guar gum**, to enhance their thickening and stabilizing effects. Locust bean gum is particularly popular in the production of **ice cream** and **dairy desserts**, where it prevents the formation of ice crystals and improves the texture of frozen products [10].
- **Gum Arabic**: Also known as **acacia gum**, gum arabic is derived from the sap of the **Acacia** tree and is primarily used in the production of **beverages**, **confectionery**, and **dairy products**. It serves as a stabilizer in emulsions and provides a smooth mouthfeel in various liquid products [11].

8.2.1.2 Pectin
Pectin is a naturally occurring polysaccharide found in the cell walls of fruits, particularly **apples** and **citrus fruits**. It is widely used in the food industry for its **gelling** properties. Pectin is most commonly used in the production of **jams**, **jellies**, and **fruit spreads**, where it helps achieve the desired consistency and spread ability. In addition to its gelling function, pectin can also be used as a **stabilizer** in beverages and

dairy products, preventing the separation of ingredients and maintaining product consistency [12].

Pectin is prized for its ability to form gels, without the need for excessive sugar, making it an ideal ingredient in **low-sugar** and **reduced-calorie** food products. Additionally, pectin has health benefits, including its ability to support digestive health due to its high fiber content, which contributes to its role in **functional foods** [13].

8.2.1.3 Starches

Starches are another class of natural thickeners used in food products. They are carbohydrates derived from plant sources such as **corn**, **potatoes**, **wheat**, and **rice**. Starches are used in the production of **soups**, **gravies**, **puddings**, **baked goods**, and **sauces**. Starches are valued for their ability to increase viscosity and improve texture while also offering cost-effective thickening solutions [14]:

- **Cornstarch**: One of the most commonly used thickeners, cornstarch is a polysaccharide derived from the **endosperm** of corn kernels. It is commonly used in sauces, gravies, and **desserts** like puddings, where it thickens when heated in the presence of liquid. It is also used in **baking** to improve the **texture** of cakes and cookies [15].
- **Potato Starch**: Derived from potatoes, potato starch is used in both **gluten-free baking** and **sauces** due to its excellent water-binding properties. It is effective at thickening soups and sauces, providing a smooth and glossy finish [16].

8.2.1.4 Seaweed Derivatives

Seaweed-derived products are another important category of natural thickeners and stabilizers. These include **agar-agar**, **carrageenan**, and **alginate**, all of which are derived from different types of seaweed [17]:

- **Agar-Agar**: Agar-agar is a gelatinous substance extracted from **red algae**. It is primarily used as a **gelling agent** in desserts, such as **jellies** and **gummy candies**. Agar has a unique ability to form firm gels even at low concentrations and remains stable at higher temperatures compared to gelatin. It is also used as a **stabilizer** in the production of dairy and nondairy alternatives [18].
- **Carrageenan**: Carrageenan is a family of **sulfated polysaccharides** extracted from red seaweeds. It is used in a variety of dairy products, including **milk**, **ice cream**, and **cheese**, where it helps to prevent crystallization, stabilize emulsions, and improve the **mouthfeel** of the products. Carrageenan is also used in **plant-based milks** as a thickening and stabilizing agent [19].
- **Alginate**: Derived from brown algae, alginate is used in food products for its ability to form **gels** and act as a **viscosity modifier**. It is commonly used in **sauces**, **dairy products**, and **beverages** to improve texture and **stability** [20].

8.2.2 Functions of Natural Thickeners and Stabilizers

Natural thickeners and stabilizers offer several critical functions in food processing.

8.2.2.1 Texture Enhancement

One of the primary functions of natural thickeners and stabilizers is to improve the texture of food products. For example, guar gum and xanthan gum are commonly added to dairy products such as yogurt and ice cream to create a creamy, smooth mouthfeel. In beverages, these ingredients help provide body and prevent products from feeling thin or watery. Pectin, on the other hand, is used to give jams, jellies, and fruit spreads the ideal consistency for spreading, while carrageenan helps improve the mouthfeel of plant-based milk alternatives [21].

8.2.2.2 Viscosity Modification

Natural thickeners increase the viscosity of liquid products without significantly altering their taste or nutritional value. Xanthan gum and guar gum are particularly effective at achieving this in low-fat or fat-free products, where they mimic the viscosity normally provided by fats. Starches, such as cornstarch and potato starch, are widely used in gravies, sauces, and soups to adjust viscosity and improve consistency, while alginate can thicken and stabilize beverages [22].

8.2.2.3 Emulsion Stabilization

Natural stabilizers play a crucial role in stabilizing emulsions – mixtures of oil and water that would otherwise separate. In the production of mayonnaise, salad dressings, and margarine, xanthan gum, guar gum, and locust bean gum are used to prevent phase separation and keep the oil and water uniformly dispersed. This helps to maintain the texture and appearance of the product during storage [23].

8.2.2.4 Shelf-Life Extension

Natural thickeners and stabilizers also help extend the shelf life of food products by preventing undesirable changes, such as phase separation, syneresis, or the formation of ice crystals. Carrageenan, for example, prevents ice crystals from forming in ice cream, ensuring a smooth, creamy texture even after freezing. Pectin is used in jams and jellies to prevent the release of water during storage, maintaining product quality and texture [24].

8.2.3 Applications of Natural Thickeners and Stabilizers in Food Products

The versatility of natural thickeners and stabilizers makes them essential ingredients in the formulation of a wide range of food products. Their ability to improve texture, stability, and consistency is invaluable in the food and beverage industry.

8.2.3.1 Dairy Products
Natural thickeners and stabilizers are extensively used in dairy products to improve texture, prevent crystallization, and stabilize emulsions. Xanthan gum and guar gum are frequently added to ice cream to enhance its smooth, creamy texture and prevent the formation of ice crystals. Carrageenan is used in milk-based beverages and plant-based milks to maintain consistency and prevent separation. Locust bean gum is used in yogurt and cheese to improve texture and prevent syneresis (the release of liquid) [25].

8.2.3.2 Beverages
In beverages, natural thickeners and stabilizers such as gum arabic, pectin, and xanthan gum help to maintain consistency, improve mouthfeel, and prevent phase separation. Guar gum and xanthan gum are used in fruit drinks and smoothies to provide thickness, while gum arabic is often employed to stabilize emulsions in carbonated soft drinks and fruit juices [26].

8.2.3.3 Confectionery
In confectionery, natural thickeners like pectin and agar-agar are used to create firm gels and chewy textures. Pectin is the primary gelling agent in fruit jams, jellies, and gummy candies, while agar-agar is a popular vegetarian alternative to gelatin in gelling products [27].

8.2.3.4 Sauces and Dressings
Natural thickeners like xanthan gum and locust bean gum are commonly used in sauces and salad dressings to modify viscosity and stabilize emulsions. These ingredients help maintain the desired thickness and consistency of sauces while also preventing the separation of oil and water phases [28].

8.2.3.5 Soups and Gravies
Starches such as cornstarch and potato starch are widely used to thicken soups, gravies, and sauces. These ingredients help improve the consistency of these products, providing a smooth and creamy texture without altering the taste or color [29].

8.3 Emulsification and Encapsulation in Beverages

8.3.1 Emulsification in Beverages

Emulsification is the process of mixing two immiscible liquids (commonly oil and water), which normally would separate over time due to differences in their polarities. In beverage formulation, emulsification is often used to stabilize the combination of water-based and oil-based ingredients, creating a homogeneous mixture. This is particularly important in drinks such as milk-based beverages, smoothies, energy drinks, and functional beverages, where oil-soluble ingredients (like flavors, vitamins, and essential oils) need to be uniformly distributed in water-based solutions [30].

8.3.2 The Role of Emulsifiers

Emulsifiers are substances that facilitate the emulsification process by reducing the surface tension between oil and water, thereby allowing them to mix more easily. In beverages, emulsifiers are often used to ensure that the oil droplets remain suspended in the water phase without separating. These emulsifiers can be **natural, synthetic**, or **semisynthetic**. **Natural emulsifiers** are derived from plant or animal sources, while synthetic emulsifiers are chemically manufactured.

Some commonly used emulsifiers in beverages include:

- **Lecithin**: A phospholipid found in egg yolks and soybeans, lecithin is one of the most commonly used natural emulsifiers. It is widely employed in beverages like **smoothies, milk drinks**, and **energy beverages** to stabilize emulsions [31].
- **Mono- and Diglycerides**: These are often used in combination with other emulsifiers in beverages to stabilize oil and water mixtures. They can be derived from **vegetable oils** and are commonly found in **beverages** like **iced tea, soft drinks**, and **dairy beverages** [32].
- **Gum Arabic**: This natural gum, extracted from acacia trees, is frequently used in beverage emulsification, especially in **fruit juices** and **flavored drinks**. It helps stabilize emulsions, particularly when dealing with **essential oils** and **Flavors** [33].
- **Xanthan Gum**: While primarily used as a **thickening agent**, xanthan gum also plays a role in **emulsifying** certain oil-based ingredients in beverages, providing additional texture and stability [34].

8.3.3 Mechanism of Emulsification

When an emulsifier is added to a mixture of oil and water, it reduces the interfacial tension between the two phases. This allows the oil droplets to break into smaller sizes and disperse evenly throughout the water phase, forming a stable emulsion. The hydrophilic (water-attracting) part of the emulsifier interacts with the water, while the lipophilic (oil-attracting) part interacts with the oil, creating a barrier that prevents the droplets from merging.

The result is a stable emulsion, which is essential for beverages that need to maintain a consistent appearance and texture over time. Without proper emulsification, beverages containing oil and water would separate, leading to undesirable sedimentation and separation, ultimately affecting product quality and consumer perception [35].

8.3.4 Types of Emulsions in Beverages

Emulsions in beverages can be broadly classified into **oil-in-water** and **water-in-oil** emulsions:

- **Oil-in-Water Emulsions**: In this type of emulsion, oil droplets are dispersed in a continuous water phase. The majority of beverages, such as **fruit juices**, **smoothies**, and **dairy-based drinks**, are oil-in-water emulsions. These emulsions are characterized by oil droplets (containing flavorings, vitamins, and essential oils) suspended in the water phase, which is the dominant phase in the formulation [36].
- **Water-in-Oil Emulsions**: Less common in beverages, water-in-oil emulsions are typically used in **fat-based beverages** like **butter** or certain **creamy alcoholic drinks**. In these emulsions, the water phase is dispersed in the oil, and the resulting texture is much thicker and creamier [36, 37].

8.3.5 Emulsification Challenges in Beverages [38]

While emulsification is essential for the creation of many beverage formulations, it can present several challenges, such as:

- **Stability**: Over time, emulsions can destabilize due to the coalescence of oil droplets, leading to **phase separation**. This is particularly true in beverages with low **viscosity** and **high fat content**.
- **Mouthfeel**: Achieving the desired sensory attributes, such as smoothness and creaminess, while maintaining emulsion stability is a delicate balance. Excessive emulsification may lead to a **greasy** or **heavy** mouthfeel, which is undesirable in many beverages.
- **Particle Size**: The size of the dispersed oil droplets affects both the **stability** and **appearance** of the beverage. Smaller droplets usually provide more stability, but they may affect the visual clarity of the drink. Proper emulsifier selection and processing conditions are necessary to achieve the ideal particle size.

8.4 Encapsulation in Beverages

Encapsulation is another advanced technique used in the formulation of beverages, particularly functional drinks, to protect sensitive ingredients, control their release, and enhance the bioavailability of nutrients. This technique involves enclosing a substance (usually a nutrient, flavor, or vitamin) within a protective coating or shell made of various materials. Encapsulation is commonly used in the beverage industry to protect ingredients from degradation, oxidation, or loss of activity during storage or processing [39].

8.4.1 Benefits of Encapsulation in Beverages

Encapsulation offers numerous advantages in beverage formulation, including:
- **Protection of Sensitive Ingredients**: Many functional ingredients, such as vitamins, probiotics, and flavors, are sensitive to heat, light, and oxygen. Encapsulation helps protect these ingredients from degradation, ensuring that they retain their efficacy and activity throughout the shelf life of the beverage [40]:
- **Controlled Release**: Encapsulation enables the controlled release of encapsulated substances over time. This is particularly useful in functional beverages that contain ingredients such as vitamins, minerals, or plant extracts that require gradual release in the digestive system for better absorption [41].
- **Enhanced Bioavailability**: Encapsulation can improve the bioavailability of certain nutrients by protecting them from digestive enzymes and promoting their absorption in the body. For example, lipophilic vitamins such as vitamin A and vitamin D can be encapsulated in lipid-based carriers to enhance their absorption in the intestines [39].
- **Improved Taste**: Encapsulation can also be used to mask the unpleasant taste of certain ingredients, such as fish oils, herbs, or plant extracts, which are often added to functional beverages. By encapsulating these ingredients, manufacturers can avoid off-flavors and improve the overall taste of the beverage [40].

8.4.2 Materials Used for Encapsulation

The materials used for encapsulating active ingredients in beverages are typically **biodegradable**, **safe**, and **food-grade**. Common encapsulating materials include:
- **Polysaccharides**: Starch, pectin, and alginate are some of the polysaccharides used for encapsulation due to their biocompatibility and ability to form gel-like structures. These materials are commonly used in beverages to encapsulate water-soluble ingredients, such as flavors and vitamins [42].

- **Proteins**: Gelatin, casein, and albumin are used as encapsulants in the beverage industry, particularly for lipophilic (oil-soluble) ingredients like essential oils and omega-3 fatty acids [43].
- **Lipids**: **Lipid-based encapsulants**, such as liposomes and solid lipid nanoparticles, are often used for encapsulating fat-soluble nutrients, such as vitamin E and omega-3 fatty acids. These lipid carriers help protect the active ingredients from oxidation and enhance their absorption in the body [44].
- **Polymers**: Polymer-based encapsulants, such as polyethylene glycol and polylactic acid, are used in certain beverage applications where controlled release and stability are required [38].

8.4.3 Encapsulation Techniques [45]

There are several techniques used for encapsulation in the beverage industry, including:
- **Spray Drying**: This is one of the most common techniques for encapsulating water-soluble ingredients in beverages. The active ingredient is dissolved in a carrier solution, which is then sprayed into hot air to evaporate the solvent and form solid particles around the active ingredient.
- **Coacervation**: This technique involves the phase separation of a polymer solution to form microcapsules around the active ingredient. Coacervation is commonly used to encapsulate **oil-soluble** substances in beverages.
- **Extrusion**: In extrusion, a substance is forced through a nozzle to create a solid shell around the active ingredient. This technique is often used for encapsulating oils and fats in beverage formulations.
- **Fluidized Bed Coating**: This technique involves the application of a coating material onto particles suspended in a fluidized bed. It is commonly used for encapsulating water-soluble ingredients, such as flavors or vitamins, in beverages.

8.4.4 Challenges in Encapsulation for Beverages [46, 47]

While encapsulation offers many advantages, it also presents certain challenges in beverage formulation:
- **Cost**: Encapsulation techniques, especially those involving lipid carriers or advanced polymers, can be expensive. This may increase the overall cost of production, making it difficult to scale for mass-market beverages.
- **Taste Masking**: While encapsulation can mask the off-flavors of certain ingredients, it can also alter the overall taste profile of the beverage. Balancing the sensory attributes of the drink while incorporating encapsulated ingredients is a delicate task.

- **Stability**: Encapsulated ingredients must remain stable over the shelf life of the beverage. Factors such as temperature, humidity, and oxygen exposure can affect the stability of encapsulated ingredients.

Emulsification and encapsulation are both essential processes in the creation of modern beverages, particularly those containing complex ingredients and multiple phases. Emulsification allows for the stable mixing of oil and water phases in beverages, while encapsulation protects sensitive ingredients and controls their release.

8.5 Use in Gluten-Free, Vegan, and Organic Food Products

8.5.1 Gluten-Free Food Products

Gluten, a protein found in wheat, barley, and rye, is responsible for giving bread dough its elasticity and structure. However, for individuals with celiac disease, non-celiac gluten sensitivity, or a wheat allergy, consuming gluten can lead to a range of health issues, including digestive discomfort, immune reactions, and inflammation. As a result, there has been an increased demand for gluten-free alternatives across the globe. This dietary shift has significantly contributed to the rise of the gluten-free food market, including gluten-free bread, pasta, baked goods, and snacks.

The challenge in formulating gluten-free products lies in replacing gluten, which gives dough its structure, elasticity, and texture. Without gluten, gluten-free products often suffer from poor texture, crumbliness, and dryness. To address this challenge, manufacturers rely on a variety of natural gums, starches, and emulsifiers that help improve texture, moisture retention, and overall product quality [48].

8.5.2 The Role of Natural Ingredients in Gluten-Free Products [49–51]

Natural gums such as guar gum, xanthan gum, locust bean gum, pectin, and various starches are essential in mimicking the structure and texture of gluten in gluten-free formulations. These ingredients play a critical role in improving the viscosity, mouthfeel, and stability of gluten-free foods, ensuring that they maintain an appealing and consistent texture throughout their shelf life.

- **Guar Gum**: Guar gum, extracted from the guar bean, is a widely used thickening and binding agent in gluten-free products. It helps provide viscosity, enhances moisture retention, and mimics the elasticity of gluten, particularly in gluten-free bread and pizza dough.
- **Xanthan Gum**: Xanthan gum is a natural polysaccharide produced by the fermentation of glucose or sucrose. It acts as an excellent emulsifier and stabilizer, improving the texture and viscosity of gluten-free batters, sauces, and dressings.

In baked goods, xanthan gum helps retain moisture and improves the product's chewiness.
- **Locust Bean Gum:** Derived from the carob tree, locust bean gum is often used in combination with other gums to enhance the texture and stability of gluten-free products. It is particularly useful in dairy-free or plant-based products, providing a creamy mouthfeel and stabilizing emulsions.
- **Rice and Potato Starch:** Both rice and potato starch are commonly used in gluten-free formulations as primary base ingredients. These starches offer a neutral flavor and contribute to the binding and moisture retention properties of gluten-free doughs and batters.
- **Pectin:** A naturally occurring carbohydrate found in fruits, pectin is widely used in gluten-free jams, jellies, and fruit-based desserts. It acts as a gelling agent, providing the desired texture and consistency in gluten-free fruit spreads and other products.

8.5.3 Challenges in Gluten-Free Food Formulation

The primary challenge in formulating gluten-free products lies in achieving a texture similar to that of gluten-containing counterparts. Gluten provides elasticity, structure, and the ability to retain moisture, all of which are vital for achieving a desirable texture in bread, cakes, and cookies. Natural gums and stabilizers can help to some extent, but getting the right balance of ingredients is crucial. Overuse of gums can lead to an overly gummy or slimy texture, while too little can result in a dry, crumbly product [50].

Additionally, gluten-free products are often more susceptible to staleness due to the lack of gluten's moisture-holding capacity. Natural ingredients such as gums and pectin can help address this issue by improving moisture retention and enhancing shelf life. However, without the correct formulation and processing conditions, gluten-free products can still lose their freshness quickly [51].

8.5.3.1 Vegan Food Products
The increasing popularity of plant-based diets and veganism, driven by ethical, environmental, and health reasons, has led to a rise in demand for vegan food products. Consumers are turning away from animal-derived products like meat, dairy, and eggs and opting for plant-based alternatives. This has prompted the food industry to innovate, creating new formulations for products like plant-based meats, dairy substitutes, and egg replacers [52].

The challenge in developing vegan products lies in replacing the functional properties of animal-derived ingredients. For example, eggs are often used for their binding, emulsifying, and gelling properties, while dairy provides creaminess, texture, and richness. To replicate these characteristics, manufacturers use a variety of natural emulsifiers, stabilizers, and thickeners derived from plants or other natural sources [53].

8.5.3.2 The Role of Natural Ingredients in Vegan Products

Natural ingredients such as gums, pectin, agar-agar, soy lecithin, and starches play an essential role in the formulation of vegan products. These ingredients help provide the necessary texture, mouthfeel, and stability that are often lost when animal-based ingredients are excluded:

– **Agar-Agar:** A popular plant-based alternative to gelatin, agar-agar is derived from seaweed and is commonly used in vegan jellies, desserts, and gummy candies. It forms a strong gel and can withstand higher temperatures than animal-based gelatin, making it ideal for vegan formulations.
– **Soy Lecithin:** Lecithin, especially soy lecithin, is a widely used emulsifier in vegan formulations. It helps blend oil-based and water-based ingredients, preventing separation and ensuring smooth texture. It is used in vegan mayo, dressings, and spreads, as well as plant-based milks and cheeses [54].
– **Xanthan Gum:** As in gluten-free formulations, xanthan gum is also commonly used in vegan food products due to its emulsifying, stabilizing, and thickening properties. It helps improve the texture and mouthfeel of vegan dairy substitutes such as vegan cheese, butter, and yogurt [55].
– **Pectin:** Pectin, derived from fruit, is commonly used in vegan products such as fruit jams, jellies, and gummy candies. It provides gelling properties, allowing the product to maintain a smooth, firm texture without the need for gelatin [56].
– **Rice Starch:** Rice starch is often used as a binding agent and thickener in vegan burgers, sausages, and other plant-based meat alternatives. It helps to provide structure and moisture retention, ensuring that the final product holds together and maintains a desirable texture [57].

8.5.3.3 Challenges in Vegan Food Formulation

Formulating vegan products often presents challenges related to texture and sensory characteristics. For example, plant-based products can sometimes have a distinct flavor that may be undesirable to some consumers. Additionally, mimicking the rich mouthfeel and creaminess of animal-based products, especially dairy, is difficult. While gums and pectin can help, achieving the perfect balance of flavor, texture, and stability requires careful ingredient selection and optimization [58].

Another challenge lies in replicating the binding and emulsifying properties of eggs. Ingredients like soy lecithin and guar gum are commonly used to replace eggs, but they may not fully replicate the exact texture and structure provided by eggs in products like cakes, mayonnaise, and pancakes [59].

8.6 Organic Food Products

Organic food products are produced using methods that avoid synthetic pesticides, fertilizers, and genetically modified organisms. The organic food industry has seen significant growth as consumers become more aware of the environmental and health impacts of conventional farming practices. Organic food is perceived as healthier and more sustainable, and it is often marketed as being free from artificial additives and preservatives [60].

8.6.1 The Role of Natural Ingredients in Organic Products

In organic food production, the use of natural ingredients such as gums, pectin, and starches is crucial. These ingredients help provide the necessary functionality texture, stability, and mouthfeel without the use of synthetic additives. Organic food formulations are typically simpler and contain fewer chemical additives, which makes the role of natural ingredients even more significant:

- **Organic Guar Gum:** Guar gum is commonly used in organic gluten-free products as a thickener and binder. It enhances the viscosity of batters and doughs and helps improve the moisture retention in gluten-free baked goods [61].
- **Organic Xanthan Gum:** Organic xanthan gum is used in organic sauces, dressings, and beverages for its emulsifying and stabilizing properties. It also contributes to the texture of dairy-free products and plant-based milk [62].
- **Organic Pectin**: Pectin is a natural carbohydrate found in fruits, and it is widely used in organic fruit-based products such as jams, jellies, and fruit spreads. It provides gelling and thickening properties, helping to achieve the desired consistency in organic products [63].
- **Organic Potato and Cornstarch**: Both organic potato starch and organic cornstarch are used as thickening agents and binding agents in organic soups, sauces, and gravies. These starches help to create a smooth texture and can be used to replace more refined flour-based thickeners [64, 65].

8.6.2 Challenges in Organic Food Formulation

One of the main challenges in organic food production is ensuring shelf stability without the use of synthetic preservatives. Organic products tend to have a shorter shelf life, which can be an issue in large-scale manufacturing. The use of natural gums, pectin, and starches helps to address this challenge by improving moisture retention and extending shelf life. However, organic products often face challenges related to texture and mouthfeel. For instance, plant-based dairy alternatives may not be as creamy as their dairy counterparts, and meat substitutes might lack the desired chewiness.

Overcoming these texture issues requires the careful selection and combination of natural ingredients to ensure the final product meets consumer expectations.

References

[1]   da Silva, D. A., Aires, G. C. M., & da Silva Pena, R. (2020). Gums – characteristics and applications in the food industry. In *Innovation in the food sector through the valorization of food and agro-food by-products*. IntechOpen.

[2]   Yemenicioğlu, A., Farris, S., Turkyilmaz, M., & Gulec, S. (2020). A review of current and future food applications of natural hydrocolloids. *International Journal of Food Science & Technology, 55*(4), 1389–1406.

[3]   Panda, H. (2010). *The Complete Book on Gums and Stabilizers for Food Industry: Gums and stabilizers Business for food industry, Gums and Stabilizers for the Food Industry, Gums and stabilizers manufacturing, Gums and stabilizers production Industry in India, Gums and stabilizers Small Business Manufacturing, Gums for food industry, How gelatin is made-production process, How to make seaweed extract, How to Manufacture Gums and Stabilizers, How to start a gums and stabilizers Production Business, How to start a successful . . . .* Asia Pacific Business Press Inc.

[4]   Sierra-López, L. D., Hernandez-Tenorio, F., Marín-Palacio, L. D., & Giraldo-Estrada, C. (2023). Coffee mucilage clarification: A promising raw material for the food industry. *Food and Humanity, 1*, 689–695.

[5]   Pegg, A. M. (2012). The application of natural hydrocolloids to foods and beverages. In *Natural food additives, ingredients and flavourings* (pp. 175–196). Woodhead Publishing.

[6]   Nussinovitch, A., Nussinovitch, M. H. A., & Hirashima, M. USING HYDROCOLLOIDS FOR THICKENING, GELLING, AND EMULSIFICATION.

[7]   Sworn, G. (2007). Natural thickeners. *Handbook of industrial water soluble polymers*, 10–31.

[8]   Mudgil, D., Barak, S., & Khatkar, B. S. (2014). Guar gum: processing, properties and food applications – a review. *Journal of food science and technology, 51*(3), 409–418.

[9]   Katzbauer, B. (1998). Properties and applications of xanthan gum. *Polymer degradation and Stability, 59*(1–3), 81–84.

[10] Barak, S., & Mudgil, D. (2014). Locust bean gum: Processing, properties and food applications – A review. *International journal of biological macromolecules, 66*, 74–80.

[11] Sulieman, A. M. E. H. (2018). Gum arabic as thickener and stabilizing agents in dairy products. *Gum arabic*, 151–165.

[12] Thakur, B. R., Singh, R. K., Handa, A. K., & Rao, M. A. (1997). Chemistry and uses of pectin – A review. *Critical Reviews in Food Science & Nutrition, 37*(1), 47–73.

[13] Khan, M., Nakkeeran, E., & Umesh-Kumar, S. (2013). Potential application of pectinase in developing functional foods. *Annual review of food science and technology, 4*(1), 21–34.

[14] Niba, L. L. (2002). Resistant starch: a potential functional food ingredient. *Nutrition & Food Science, 32*(2), 62–67.

[15] Heydari, A., Alemzadeh, I., & Vossoughi, M. (2013). Functional properties of biodegradable corn starch nanocomposites for food packaging applications. *Materials & Design, 50*, 954–961.

[16] Xu, J., Li, Y., Kaur, L., Singh, J., & Zeng, F. (2023). Functional food based on Potato. *Foods, 12*(11), 2145.

[17] Mohamed, S., Hashim, S. N., & Rahman, H. A. (2012). Seaweeds: A sustainable functional food for complementary and alternative therapy. *Trends in food science & technology, 23*(2), 83–96.

[18] Pandya, Y. H., Bakshi, M., Sharma, A., Pandya, Y. H., & Pandya, H. (2022). Agar-agar extraction, structural properties and applications: A review. *The Pharma Innovation Journal, 6*, 1151–1157.

[19]   Prasetyaningrum, A., & Praptyana, I. R. (2019, June). Carrageenan: nutraceutical and functional food as future food. In *IOP Conference Series: Earth and Environmental Science* (Vol. 292, No. 1, p. 012068). IOP Publishing.

[20]   Qin, Y., Jiang, J., Zhao, L., Zhang, J., & Wang, F. (2018). Applications of alginate as a functional food ingredient. In *Biopolymers for food design* (pp. 409–429). Academic Press.

[21]   Ribes, S., Grau, R., & Talens, P. (2022). Use of chia seed mucilage as a texturing agent: Effect on instrumental and sensory properties of texture-modified soups. *Food Hydrocolloids, 123*, 107171.

[22]   Garcia-Barradas, O., Esteban-Cortina, A., Mendoza-Lopez, M. R., Ortiz-Basurto, R. I., Díaz-Ramos, D. I., & Jimenez-Fernandez, M. (2023). Chemical modification of Opuntia ficus-indica mucilage: characterization, physicochemical, and functional properties. *Polymer Bulletin, 80*(8), 8783–8798.

[23]   Cárdenas, A., Higuera-Ciapara, I., & Goycoolea, F. M. (1997). Rheology and aggregation of cactus (Opuntia ficus-indica) mucilage in solution. *Journal of the Professional Association for cactus development, 2*, 152–159.

[24]   Behbahani, B. A., Shahidi, F., Yazdi, F. T., Mortazavi, S. A., & Mohebbi, M. (2017). Use of Plantago major seed mucilage as a novel edible coating incorporated with Anethum graveolens essential oil on shelf life extension of beef in refrigerated storage. *International journal of biological macromolecules, 94*, 515–526.

[25]   Ribes, S., Peña, N., Fuentes, A., Talens, P., & Barat, J. M. (2021). Chia (Salvia hispanica L.) seed mucilage as a fat replacer in yogurts: Effect on their nutritional, technological, and sensory properties. *Journal of Dairy Science, 104*(3), 2822–2833.

[26]   KC, Y., Subba, R., Shiwakoti, L. D., Dhungana, P. K., Bajagain, R., Chaudhary, D. K., . . . & Dahal, R. H. (2021). Utilizing coffee pulp and mucilage for producing alcohol-based beverage. *Fermentation, 7*(2), 53.

[27]   Du Toit, L. (2018). *Celling properties of cactus pear mucilage-hydrocolloid combinations in a sugar-based confectionery* (Doctoral dissertation, University of the Free State).

[28]   Campbell, B. (2019). Current emulsifier trends in dressings and sauces. In *Food Emulsifiers and Their Applications* (pp. 285–298). Cham: Springer International Publishing.

[29]   Ribes, S., Grau, R., & Talens, P. (2022). Use of chia seed mucilage as a texturing agent: Effect on instrumental and sensory properties of texture-modified soups. *Food Hydrocolloids, 123*, 107171.

[30]   Waghmare, R., R, P., Moses, J. A., & Anandharamakrishnan, C. (2022). Mucilages: Sources, extraction methods, and characteristics for their use as encapsulation agents. *Critical Reviews in Food Science and Nutrition, 62*(15), 4186–4207.

[31]   Knott, M., Ani, M., Kroener, E., & Diehl, D. (2022). Effect of environmental conditions on physical properties of maize root mucilage.

[32]   Cenobio-Galindo, A. D. J., Díaz-Monroy, G., Medina-Pérez, G., Franco-Fernández, M. J., Ludeña-Urquizo, F. E., Vieyra-Alberto, R., & Campos-Montiel, R. G. (2019). Multiple emulsions with extracts of cactus pear added in a yogurt: Antioxidant activity, in vitro simulated digestion and shelf life. *Foods, 8*(10), 429.

[33]   McNamee, B. F., O'Riorda, E. D., & O'Sullivan, M. (1998). Emulsification and microencapsulation properties of gum arabic. *Journal of Agricultural and Food Chemistry, 46*(11), 4551–4555.

[34]   Desplanques, S., Renou, F., Grisel, M., & Malhiac, C. (2012). Impact of chemical composition of xanthan and acacia gums on the emulsification and stability of oil-in-water emulsions. *Food Hydrocolloids, 27*(2), 401–410.

[35]   Lee, L. L., Niknafs, N., Hancocks, R. D., & Norton, I. T. (2013). Emulsification: mechanistic understanding. *Trends in Food Science & Technology, 31*(1), 72–78.

[36]   Wong, S. F., Lim, J. S., & Dol, S. S. (2015). Crude oil emulsion: A review on formation, classification and stability of water-in-oil emulsions. *Journal of Petroleum Science and Engineering, 135*, 498–504.

[37]   Roland, I., Piel, G., Delattre, L., & Evrard, B. (2003). Systematic characterization of oil-in-water emulsions for formulation design. *International journal of pharmaceutics, 263*(1–2), 85–94.

[38] Stephen, A. M., & Churms, S. C. (1995). Gums and mucilages. *FOOD SCIENCE AND TECHNOLOGY-NEW YORK-MARCEL DEKKER-*, 377–377.

[39] Waghmare, R., R, P., Moses, J. A., & Anandharamakrishnan, C. (2022). Mucilages: Sources, extraction methods, and characteristics for their use as encapsulation agents. *Critical Reviews in Food Science and Nutrition*, *62*(15), 4186–4207.

[40] Given Jr, P. S. (2009). Encapsulation of flavors in emulsions for beverages. *Current Opinion in Colloid & Interface Science*, *14*(1), 43–47.

[41] Given Jr, P. S. (2009). Encapsulation of flavors in emulsions for beverages. *Current Opinion in Colloid & Interface Science*, *14*(1), 43–47.

[42] Devi, N., Sarmah, M., Khatun, B., & Maji, T. K. (2017). Encapsulation of active ingredients in polysaccharide–protein complex coacervates. *Advances in colloid and interface science*, *239*, 136–145.

[43] Ye, C., & Chi, H. (2018). A review of recent progress in drug and protein encapsulation: Approaches, applications and challenges. *Materials Science and Engineering: C*, *83*, 233–246.

[44] Cimino, C., Maurel, O. M., Musumeci, T., Bonaccorso, A., Drago, F., Souto, E. M. B., . . . & Carbone, C. (2021). Essential oils: Pharmaceutical applications and encapsulation strategies into lipid-based delivery systems. *Pharmaceutics*, *13*(3), 327.

[45] Abd El-Kader, A., & Abu Hashish, H. (2020). Encapsulation techniques of food bioproduct. *Egyptian Journal of Chemistry*, *63*(5), 1881–1909.

[46] Abbas, M. S., Saeed, F., Afzaal, M., Jianfeng, L., Hussain, M., Ikram, A., & Jabeen, A. (2022). Recent trends in encapsulation of probiotics in dairy and beverage: A review. *Journal of Food Processing and Preservation*, *46*(7), e16689.

[47] Maswal, M., & Dar, A. A. (2014). Formulation challenges in encapsulation and delivery of citral for improved food quality. *Food Hydrocolloids*, *37*, 182–195.

[48] Banu, Z., Somraj, B., Geetha, A., Khan, A. A., Srinu, B., Vamshi, J., . . . & Saidaiah, P. (2024). A comprehensive review of biochemical, functional, and dietary implications of gluten. *Annals of Phytomedicine*, *13*(1), 196–208.

[49] Pellegrini, N., & Agostoni, C. (2015). Nutritional aspects of gluten-free products. *Journal of the Science of Food and Agriculture*, *95*(12), 2380–2385.

[50] Šmídová, Z., & Rysová, J. (2022). Gluten-free bread and bakery products technology. *Foods*, *11*(3), 480.

[51] Difonzo, G., de Gennaro, G., Pasqualone, A., & Caponio, F. (2022). Potential use of plant-based by-products and waste to improve the quality of gluten-free foods. *Journal of the Science of Food and Agriculture*, *102*(6), 2199–2211.

[52] Bärebring, L., Lamberg-Allardt, C., Thorisdottir, B., Ramel, A., Söderlund, F., Arnesen, E. K., . . . & Åkesson, A. (2023). Intake of vitamin B12 in relation to vitamin B12 status in groups susceptible to deficiency: a systematic review. *Food & Nutrition Research*, *67*, 10–29219.

[53] Chavan, J. K., Kadam, S. S., & Beuchat, L. R. (1989). Nutritional improvement of cereals by fermentation. *Critical Reviews in Food Science & Nutrition*, *28*(5), 349–400.

[54] Ścieszka, S., & Klewicka, E. (2019). Algae in food: A general review. *Critical reviews in food science and nutrition*, *59*(21), 3538–3547.

[55] Asase, R. V., & Glukhareva, T. V. (2024). Production and application of xanthan gum – prospects in the dairy and plant-based milk food industry: a review. *Food Science and Biotechnology*, *33*(4), 749–767.

[56] Yemenicioğlu, A., Farris, S., Turkyilmaz, M., & Gulec, S. (2020). A review of current and future food applications of natural hydrocolloids. *International Journal of Food Science & Technology*, *55*(4), 1389–1406.

[57] Yano, H., & Fu, W. (2022). Effective use of plant proteins for the development of "new" foods. *Foods*, *11*(9), 1185.

[58] Marques-Lopes, I., Martínez-Biarge, M., Martínez-Pineda, M., & Menal-Puey, S. (2025). Optimizing Vegan Nutrition: Current Challenges and Potential Solutions. *Applied Sciences*, *15*(17), 9485.

[59] Alcorta, A., Porta, A., Tárrega, A., Alvarez, M. D., & Vaquero, M. P. (2021). Foods for plant-based diets: Challenges and innovations. *Foods*, *10*(2), 293.

[60] Magkos, F., Arvaniti, F., & Zampelas, A. (2003). Organic food: nutritious food or food for thought? A review of the evidence. *International journal of food sciences and nutrition*, *54*(5), 357–371.

[61] Thombare, N., Jha, U., Mishra, S., & Siddiqui, M. Z. (2016). Guar gum as a promising starting material for diverse applications: A review. *International journal of biological macromolecules*, *88*, 361–372.

[62] Berninger, T., Dietz, N., & González López, Ó. (2021). Water-soluble polymers in agriculture: xanthan gum as eco-friendly alternative to synthetics. *Microbial Biotechnology*, *14*(5), 1881–1896.

[63] Mellinas, C., Ramos, M., Jiménez, A., & Garrigós, M. C. (2020). Recent trends in the use of pectin from agro-waste residues as a natural-based biopolymer for food packaging applications. *Materials*, *13*(3), 673.

[64] Djaman, K., Sanogo, S., Koudahe, K., Allen, S., Saibou, A., & Essah, S. (2021). Characteristics of organically grown compared to conventionally grown potato and the processed products: A review. *Sustainability*, *13*(11), 6289.

[65] Santana, Á. L., & Meireles, M. A. A. (2014). New starches are the trend for industry applications: a review. *Food public health*, *4*(5), 229–241.

9 Agricultural Applications

9.1 Introduction

Mucilage and gums play a vital role in sustainable agriculture, particularly as components of biofertilizers and soil conditioners [1]. These natural polysaccharides improve soil texture and water retention, which are crucial for plant growth, especially in drought-prone areas. When used in biofertilizers, mucilage and gums serve as effective carriers for beneficial microbes, enhancing their stability and ensuring their slow and targeted release to plant roots [2]. This controlled delivery boosts nutrient availability and uptake, contributing to healthier crops. As soil conditioners, they improve soil structure, increase porosity, and reduce compaction, leading to better root penetration and aeration [3]. Their ability to hold moisture near the root zone reduces the need for frequent irrigation, conserving water resources. Applications of mucilage and gums in sustainable agriculture also include seed coating, where they act as natural binders to protect seeds and promote uniform germination. Through these mechanisms, they contribute significantly to enhancing crop yield and soil health without relying on synthetic inputs. Being biodegradable and nontoxic, these natural substances align with eco-friendly farming practices and help reduce environmental pollution. Therefore, the integration of mucilage and gums in agricultural practices supports long-term soil fertility, plant productivity, and environmental sustainability [4].

9.2 Role of Mucilage and Gums in Biofertilizers

Biofertilizers are designed to introduce beneficial microorganisms into the soil to promote plant growth through mechanisms like nitrogen fixation, phosphorus solubilization, and growth hormone production [2]. Mucilage and gums play a critical role in enhancing the effectiveness of biofertilizers by providing essential support to the microorganisms that work to benefit the plants. The multifaceted contributions of mucilage and gums are outlined as follows:

9.2.1 Microbial Growth Enhancement

Mucilage and gums serve as nutrient-rich substrates that provide an ideal carbon source for microorganisms in the soil. The presence of these polysaccharides allows beneficial microbes – such as nitrogen-fixing bacteria, fungi, and actinomycetes to thrive, as they are crucial for processes like nitrogen fixation and organic matter decomposition [5]. These microorganisms are responsible for converting atmospheric nitrogen into a form that plants can absorb, which is essential for plant growth. The

https://doi.org/10.1515/9783111673509-009

carbon present in mucilage and gums fuels the microbial metabolism, promoting their proliferation and activity. This, in turn, increases the biological processes in the soil, such as nutrient cycling, which improves the fertility of the soil.

Furthermore, mucilage and gums support the establishment of a diverse microbial community, which is vital for maintaining a healthy soil ecosystem. The increased microbial growth enhances nutrient availability, reduces soil-borne pathogens, and improves soil health, ultimately contributing to better plant growth and higher yields [6].

9.2.2 Improved Adhesion of Microorganisms

One of the key challenges faced by biofertilizers is ensuring that beneficial microorganisms remain attached to plant roots, where they can exert their positive effects. Mucilage and gums provide a gel-like, sticky texture that enhances the adhesion of microorganisms to plant roots and soil particles. This improves the effectiveness of rhizobacteria and other beneficial microbes, which assist in nutrient uptake, disease resistance, and overall plant growth [7].

The adhesive properties of mucilage and gums ensure that microorganisms stay close to the root zone, where they can form beneficial biofilms, which further protect the microorganisms from desiccation, predation, and other stressors. These biofilms also increase the efficiency of nutrient exchange between the plant and the microorganisms, creating a more symbiotic relationship that benefits both parties. The adhesion of microorganisms to plant roots helps in the formation of the rhizosphere, a critical area where beneficial soil microbes interact with plants to promote growth and nutrient uptake [8].

9.2.3 Slow Release of Nutrients

Mucilage and gums are not only beneficial for microbial growth but also serve as controlled nutrient carriers. They can slowly release essential nutrients like nitrogen, phosphorus, and potassium over time, preventing their leaching and ensuring that plants receive a steady supply of nutrients. This slow-release mechanism is particularly advantageous in biofertilizers, as it reduces the need for frequent applications and minimizes nutrient wastage [9].

In regions where rainfall is erratic or where soils are prone to nutrient leaching (such as sandy soils), mucilage and gums act as a buffer, holding nutrients in place and releasing them gradually to the plants. This controlled nutrient delivery helps plants maintain steady growth, even under suboptimal conditions, and enhances the sustainability of agricultural systems by reducing the dependence on synthetic fertilizers [10].

9.2.4 Protection of Microorganisms

Environmental stressors such as temperature fluctuations, drought, and high salinity can significantly reduce the viability of microorganisms in biofertilizers. Mucilage and gums provide an extra layer of protection to these microorganisms, shielding them from extreme conditions and enhancing their survival rate in the soil. By encapsulating microorganisms in a gel-like structure, mucilage and gums prevent desiccation and help maintain moisture levels around microbial populations, ensuring their longevity.

Moreover, the viscosity and chemical properties of mucilage and gums provide a physical barrier that reduces microbial exposure to harmful chemicals and toxic substances in the soil. This protection extends the activity of biofertilizers, allowing them to work more effectively over a longer period, thus enhancing soil fertility and improving plant health [9, 10].

9.3 Role of Mucilage and Gums in Soil Conditioners

Soil conditioners are materials that improve the physical properties of soil, such as texture, water retention, and aeration, making them essential for maintaining soil health and productivity. Mucilage and gums contribute to these processes by interacting with soil particles and promoting the formation of aggregates. The role of mucilage and gums in soil conditioning is discussed below:

9.3.1 Improved Water Retention

One of the most significant contributions of mucilage and gums to soil conditioners is their ability to enhance water retention. Mucilage, in particular, has a high capacity for absorbing and retaining water, making it highly beneficial in arid or drought-prone regions. This water-holding capacity helps improve soil moisture availability, ensuring that plants have access to water during dry spells.

In addition to water retention, mucilage and gums also regulate the rate at which water is released to the plants. This slow release ensures that the plants can access water over an extended period, reducing the need for frequent irrigation. As a result, mucilage and gums contribute to more water-efficient agricultural practices, which is particularly important in regions where water resources are limited [11].

9.3.2 Soil Aggregation and Structure

Mucilage and gums are effective in improving soil aggregation. Their viscous properties allow them to bind soil particles together, forming aggregates that improve soil structure. Soil aggregation is crucial for maintaining optimal porosity, reducing compaction, and enhancing root penetration. These improvements make the soil more aerated, which in turn promotes healthy root growth and better oxygen exchange in the soil [3].

The formation of aggregates also reduces the risk of soil erosion. When soil particles are bound together by mucilage and gums, they are less likely to be carried away by wind or water, thus preserving the topsoil. In this way, mucilage and gums contribute to the long-term sustainability of agricultural systems by preventing the loss of valuable soil resources [12].

9.3.3 Erosion Control

Erosion is a significant problem in many parts of the world, especially in regions with loose, sandy soils or areas prone to heavy rainfall. The adhesive properties of mucilage and gums help bind the soil particles together, reducing the risk of soil erosion. This is particularly important in areas where maintaining the topsoil is crucial for sustainable agriculture and food production.

By acting as a natural glue that holds soil particles together, mucilage and gums can significantly reduce wind and water erosion. This property makes them highly valuable in erosion control programs, helping to maintain soil fertility and prevent the loss of vital agricultural land [13].

9.3.4 pH Buffering

Soil pH plays a critical role in the availability of nutrients to plants and the activity of soil microorganisms. Mucilage has been shown to have pH-buffering properties, helping to maintain a stable pH environment in the soil. This is particularly beneficial in preventing soil acidification or alkalinization, which can have detrimental effects on plant growth.

By stabilizing soil pH, mucilage and gums contribute to a more conducive environment for both plants and microorganisms. This is particularly important in areas where the soil pH fluctuates due to factors like irrigation or the application of fertilizers [14].

9.3.5 Enhanced Nutrient Availability

Mucilage and gums can also improve nutrient availability in the soil by reducing nutrient leaching. In soils that are prone to leaching, particularly sandy soils, nutrients are often lost to groundwater before plants can absorb them. Mucilage and gums help retain these nutrients in the soil by binding to them, preventing them from being washed away by rain or irrigation. This process ensures that nutrients remain available to plants for a more extended period, improving nutrient use efficiency.

Furthermore, mucilage and gums also enhance the availability of micronutrients, which are often difficult for plants to access in the soil. By binding to essential minerals and trace elements, mucilage and gums make these nutrients more accessible to plants, thus promoting better growth and higher yields [15].

9.4 Synergistic Effects in Biofertilizers and Soil Conditioners

When used together, mucilage and gums can create synergistic effects that enhance the overall performance of both biofertilizers and soil conditioners. Some of the most important synergistic effects include:

9.4.1 Increased Microbial Efficiency

By providing a supportive environment for beneficial microorganisms, mucilage and gums can increase the overall efficiency of biofertilizers. The enhanced adhesion, protection from environmental stress, and nutrient availability provided by mucilage and gums allow microorganisms to thrive and function optimally. This results in improved nutrient uptake by plants, increased growth, and better yields [16].

9.4.2 Improved Soil Fertility and Structure

When mucilage and gums are used as soil conditioners, they improve both the physical structure of the soil and its microbial health. The formation of soil aggregates, enhanced water retention, and better aeration lead to improved soil conditions for plant growth. At the same time, the improved microbial activity due to the presence of mucilage and gums further enhances soil fertility, creating a positive feedback loop that sustains soil health over time [17].

9.4.3 Sustainable Agriculture

The use of natural substances like mucilage and gums reduces the need for synthetic chemicals such as fertilizers and pesticides. By improving soil fertility and promoting plant health through biological processes, mucilage and gums support more sustainable agricultural practices. These substances contribute to the long-term sustainability of farming systems by improving soil health, reducing chemical inputs, and enhancing crop yields in a more environmentally friendly manner.

Mucilage and gums are essential natural substances that play a multifaceted role in both biofertilizers and soil conditioners [12].

9.5 Applications in Sustainable Agriculture

Sustainable agriculture is critical for the long-term health of the planet and its ecosystems.

9.5.1 Water Conservation and Drought Management

Water scarcity is one of the most pressing challenges in modern agriculture, especially in arid and semiarid regions. Mucilage and gums play a significant role in addressing this challenge by enhancing water retention in the soil, thereby reducing the need for frequent irrigation and ensuring that plants receive a consistent supply of moisture during dry periods [18].

9.5.2 Water Retention

Mucilage and gums have a remarkable ability to absorb and retain water many times their weight, which is crucial for maintaining soil moisture in areas with limited water availability. These natural substances help form a gel-like structure in the soil that acts like a sponge, slowly releasing water to plants as needed. This capacity for water retention can significantly improve the moisture-holding capacity of soils, especially in sandy soils that typically have poor water retention capabilities. In dryland farming systems, mucilage and gums act as water-storing agents that help reduce evaporation losses, thus ensuring that crops receive adequate moisture even during periods of drought [19].

9.5.3 Reduced Irrigation Needs

The water-retaining properties of mucilage and gums reduce the frequency of irrigation, making them invaluable in areas where water is scarce. By slowing down the rate at which water evaporates or leaches from the soil, mucilage and gums ensure that crops can access water for a longer period without the need for frequent irrigation cycles. This not only saves water but also reduces the overall energy consumption associated with irrigation. In turn, farmers can save on water costs and minimize the carbon footprint of agricultural operations [20].

9.5.4 Protection During Drought

Mucilage and gums can serve as a protective shield for plants during drought conditions. By improving the water-holding capacity of the soil, they help ensure that crops have access to sufficient moisture even during prolonged dry spells. In regions where droughts are frequent and intense, the application of mucilage and gums in the soil can act as a drought mitigation strategy, enabling farmers to continue cultivating crops even in challenging environmental conditions [21].

9.5.5 Soil Health and Fertility

Soil health is the foundation of sustainable agriculture. Healthy soils foster better crop growth, support biodiversity, and promote efficient nutrient cycling. Mucilage and gums play a vital role in improving soil health by enhancing soil structure, supporting beneficial microbial activity, and preventing soil erosion [11].

9.5.5.1 Improved Soil Structure

Soil structure refers to the arrangement of soil particles into aggregates, which is essential for maintaining soil porosity, aeration, and water infiltration. Mucilage and gums act as natural soil binders, helping to aggregate soil particles and form larger clumps or aggregates. These aggregates improve the physical structure of the soil by reducing compaction, increasing pore space, and promoting better water and air flow through the soil. The improved structure encourages healthy root development, which is essential for optimal plant growth. In addition, better soil aggregation reduces surface crusting, which can impede water infiltration and cause runoff [1, 13].

The aggregation of soil particles also prevents the formation of hardpan layers that restrict root penetration. This enhanced soil structure leads to more efficient water retention, better root growth, and greater microbial activity, which are key elements for long-term soil fertility [19].

9.5.5.2 Enhanced Microbial Activity
Soil microorganisms, including bacteria, fungi, and actinomycetes, are essential for nutrient cycling and the breakdown of organic matter. Mucilage and gums provide an ideal environment for these microorganisms by acting as a nutrient-rich substrate. The polysaccharides in mucilage and gums serve as a food source for beneficial microbes, supporting their growth and activity. This enhanced microbial activity accelerates the decomposition of organic matter and increases the availability of essential nutrients like nitrogen, phosphorus, and sulfur to plants [17].

The increased microbial activity in the soil also helps maintain soil health by promoting natural nutrient cycling. The presence of mucilage and gums in the soil helps create a more balanced microbial ecosystem, which in turn supports the sustainability of farming systems by reducing the need for chemical fertilizers. By enhancing the microbial population, mucilage and gums contribute to the natural processes that maintain soil fertility, improving the efficiency of nutrient use in crops [21].

9.5.5.3 Erosion Control
Soil erosion is a significant threat to soil fertility, particularly in regions with high rainfall or strong winds. Erosion removes the topsoil, which is rich in nutrients and organic matter, leaving the land less fertile and harder to manage. Mucilage's sticky properties help bind soil particles together, preventing them from being detached and carried away by wind or water. By stabilizing the soil, mucilage and gums reduce the risk of erosion and protect valuable topsoil. This is particularly important in conservative agriculture practices, where preserving soil structure and fertility is a priority [22].

Mucilage and gums can be applied as a soil conditioner in erosion-prone areas to prevent soil loss and ensure long-term soil fertility. By enhancing the soil's resistance to erosion, these natural compounds contribute to the sustainability of agricultural systems, ensuring that farmers can maintain productive land over time [3, 22].

9.5.6 Reduction in Chemical Inputs

One of the most compelling benefits of using mucilage and gums in sustainable agriculture is their ability to reduce the need for chemical fertilizers and pesticides. By enhancing soil fertility and promoting healthy microbial activity, mucilage and gums provide natural alternatives to synthetic inputs, reducing the environmental impact of farming [23].

9.5.6.1 Lower Chemical Fertilizer Dependence
Chemical fertilizers, while effective in promoting plant growth, come with several drawbacks, including soil degradation, nutrient imbalances, and environmental pollu-

tion. Mucilage and gums, by improving soil health and enhancing microbial activity, can reduce the reliance on synthetic fertilizers. The increased microbial activity in the soil promotes the natural availability pf nutrients, such as nitrogen, phosphorus, and potassium, making them more accessible to plants [24].

The slow-release properties of mucilage and gums further enhance nutrient efficiency by reducing nutrient leaching and promoting a steady supply of nutrients over time. This controlled nutrient release minimizes the need for frequent fertilization and reduces the risk of nutrient runoff into water bodies. As a result, mucilage and gums contribute to more sustainable nutrient management, improving soil fertility without the harmful effects of chemical fertilizers [25].

9.5.6.2 Pesticide Reduction

The application of mucilage and gums also contributes to pest and disease management by fostering a healthy and diverse microbial ecosystem in the soil. Beneficial soil microorganisms, such as fungi and bacteria, can help control pests and diseases by outcompeting harmful pathogens for resources and space. In this way, mucilage and gums indirectly support natural pest control, reducing the need for chemical pesticides.

By promoting soil biodiversity, mucilage and gums contribute to biological pest control, which is both effective and environmentally friendly. The reduced reliance on chemical pesticides helps preserve nontarget organisms, such as beneficial insects and pollinators, and reduces the impact of pesticides on surrounding ecosystems [26].

9.5.7 Climate Change Mitigation

Agriculture is both a contributor to and a victim of climate change. On the one hand, agricultural practices can release significant amounts of greenhouse gases, such as carbon dioxide, methane, and nitrous oxide, into the atmosphere. On the other hand, climate change exacerbates challenges like drought, flooding, and temperature extremes that impact crop productivity. Mucilage and gums can help mitigate some of these effects, making agriculture more resilient and reducing its environmental impact [27].

9.5.7.1 Carbon Sequestration

Mucilage and gums contribute to soil carbon sequestration by enhancing soil aggregation. Soil aggregation improves the storage of organic carbon in the form of stable carbon compounds, which helps mitigate climate change by trapping carbon in the soil. By increasing soil organic matter, mucilage and gums promote long-term carbon storage, reducing the amount of carbon dioxide released into the atmosphere [28]. This process of carbon sequestration can significantly reduce the carbon footprint of agricultural operations and make farming systems more resilient to the impacts of climate change.

9.5.7.2 Soil Resilience to Climate Extremes
Mucilage and gums also enhance the resilience of soils to extreme weather events, such as heavy rainfall, drought, and temperature fluctuations. By improving water retention, soil structure, and microbial activity, these natural compounds make soils more adaptable to changing climate conditions. In regions prone to climate extremes, mucilage and gums help ensure that crops can thrive despite fluctuations in weather patterns, leading to more stable and predictable yields [26].

9.5.8 Sustainable Pest and Disease Management

The growing reliance on chemical pesticides has led to several environmental and health concerns, including pesticide resistance, contamination of water sources, and the destruction of beneficial organisms in the soil [29]. Mucilage and gums provide a natural alternative to chemical pesticides by supporting the growth of beneficial microorganisms that help control pests and diseases.

9.5.8.1 Beneficial Microorganism Support
Mucilage and gums create an environment conducive for beneficial soil microbes, such as nitrogen-fixing bacteria, mycorrhizal fungi, and predatory microorganisms. These organisms help control pests by outcompeting harmful pathogens for nutrients and space. In addition, some soil microbes produce natural pesticides, such as antifungal compounds or antibacterial substances, which help protect plants from diseases [28].

9.5.8.2 Reduced Use of Pesticides
By supporting a healthy and diverse microbial ecosystem, mucilage and gums reduce the need for synthetic pesticides. This not only benefits the environment but also helps farmers lower their input costs and avoid the harmful effects of chemical pesticides on human health and nontarget species. The use of mucilage and gums aligns with integrated pest management (IPM) practices, where the goal is to control pests through a combination of biological, physical, and chemical methods in an environmentally responsible manner [27].

9.5.9 Improved Crop Yield and Quality

Sustainable agriculture aims not only to minimize environmental impact but also to improve the efficiency and productivity of farming operations. Mucilage and gums contribute to both of these goals by enhancing soil health, improving water retention, and supporting nutrient uptake [22].

9.5.9.1 Enhanced Nutrient Uptake
The improved soil structure and water retention provided by mucilage and gums lead to healthier and more robust root systems. With better root development, plants can access nutrients more effectively from the soil. The enhanced microbial activity facilitated by mucilage and gums also plays a crucial role in mobilizing nutrients, ensuring that plants receive a balanced supply of essential minerals and nutrients [24].

9.5.9.2 Increased Crop Productivity
By improving soil fertility, supporting water retention, and promoting healthy root growth, mucilage and gums help increase crop yields in a sustainable manner. This allows farmers to produce more food with fewer inputs, improving productivity while reducing the environmental impact of farming. Moreover, mucilage and gums enhance crop quality by promoting better nutrient uptake and reducing plant stress, resulting in healthier, more resilient crops [28].

9.5.10 Eco-friendly and Cost-Effective

The use of mucilage and gums in agriculture is not only beneficial for the environment but also for the farmers' bottom line [30]. These natural substances are biodegradable and have minimal environmental impact compared to synthetic fertilizers and chemicals. Additionally, mucilage and gums are often more affordable than chemical inputs, making them an attractive option for farmers seeking to reduce their input costs [31].

9.5.10.1 Biodegradability
Since mucilage and gums are derived from natural plant sources, they are biodegradable and have a minimal environmental footprint. This makes them a safer alternative to synthetic chemicals, which can persist in the environment for years and cause long-term damage to ecosystems [30].

9.5.10.2 Low-Cost Inputs

Mucilage and gums, being plant-based, are often more affordable than synthetic fertilizers and soil amendments. Their widespread availability and low-cost production make them an economically viable option for farmers, particularly those in developing countries or small-scale farming operations. By using mucilage and gums, farmers can reduce their reliance on expensive synthetic inputs and improve the economic sustainability of their operations [30].

9.6 Applications in Various Farming Systems

Mucilage and gums can be used in a wide range of farming systems, from traditional agriculture to more specialized practices like agroforestry, permaculture, and conservation tillage.

9.6.1 Agroforestry

In agroforestry systems, mucilage and gums can be used to improve soil health under tree canopies, increase water retention in mixed cropping systems, and support the growth of understory crops. The ability of mucilage and gums to improve soil structure and water retention makes them ideal for supporting diverse plant species in agroforestry systems, which aim to integrate trees, crops, and livestock in a sustainable manner [32].

9.6.2 Permaculture

Permaculture systems, which focus on sustainability and the use of natural resources, can benefit greatly from the use of mucilage and gums. These compounds help improve soil fertility, conserve water, and enhance biodiversity, all of which are key principles of permaculture. By supporting healthy soil ecosystems, mucilage and gums contribute to the long-term success of permaculture farms and gardens [33].

9.6.3 Conservation Tillage

In reduced tillage systems, where soil disturbance is minimized, mucilage and gums help maintain soil structure and microbial health. These systems aim to preserve soil fertility by reducing the physical disruption of the soil, and mucilage and gums play a critical role in ensuring that the soil remains productive over time. Their ability to enhance soil aggregation and microbial activity makes them a valuable addition to conservation tillage practices [34].

9.6.4 Enhancing Crop Yield and Soil Health

Central to this challenge is the need to enhance crop yield while maintaining or improving soil health. Soil degradation, water scarcity, and overreliance on synthetic fertilizers and pesticides threaten the sustainability of agricultural systems. In this context, natural substances like mucilage and gums have emerged as potent tools for improving soil structure, boosting crop productivity, and reducing environmental impact [35].

9.6.5 Improving Soil Structure and Fertility

Soil structure plays a pivotal role in determining how well soil retains water, supports root growth, and facilitates nutrient uptake. The addition of mucilage and gums to soil can significantly improve its structure and overall fertility, making it more conducive to plant growth [28].

9.6.5.1 Soil Aggregation and Texture

Mucilage and gums are natural soil binders that help aggregate soil particles into larger clumps or aggregates. These aggregates improve soil texture, making it more resistant to compaction. Compacted soils often suffer from poor water infiltration and reduced oxygen availability, both of which limit root growth and nutrient uptake. By reducing soil compaction, mucilage and gums enhance soil porosity, facilitating better root penetration and improving soil aeration.

Aggregated soil also allows for better water movement through the soil profile, which is essential for proper hydration of plant roots. As a result, mucilage and gums can improve the physical properties of the soil, ensuring that plants have access to a more favorable growing environment [29, 30].

9.6.5.2 Increased Soil Porosity

The formation of aggregates leads to increased pore spaces in the soil, which improves aeration. This is crucial because soil aeration facilitates the flow of oxygen to plant roots and beneficial soil organisms. It also helps reduce the risk of waterlogging in soils that tend to retain excess moisture. Increased porosity not only promotes root expansion but also supports the thriving of beneficial soil microorganisms that are essential for maintaining soil health.

Studies have shown that the addition of natural substances like mucilage to soils can increase the total porosity of soils by up to 20%, significantly improving both water infiltration and oxygen exchange, both of which are critical for healthy crop development [26–28].

9.6.5.3 Boosting Nutrient Availability and Uptake

Soil fertility is influenced not only by the physical properties of the soil but also by its ability to supply nutrients to plants. Mucilage and gums contribute to enhanced nutrient availability and uptake by promoting nutrient cycling, improving nutrient release rates, and supporting beneficial microbial activity [32].

9.6.5.4 Improved Nutrient Cycling

Mucilage and gums act as carriers for essential plant nutrients such as nitrogen, phosphorus, and potassium. These nutrients are released gradually and steadily, reducing the risk of nutrient leaching, especially in sandy or highly permeable soils. This slow release ensures that nutrients are available to plants over an extended period, thereby reducing the need for frequent fertilizer applications.

By supporting the slow and controlled release of nutrients, mucilage and gums help optimize nutrient availability throughout the growing season, ensuring that crops receive a constant supply of essential elements for growth. The controlled release also minimizes the risk of nutrient runoff into nearby water bodies, which can contribute to environmental pollution [30].

9.6.5.5 Nutrient Release and Availability

In addition to promoting nutrient cycling, mucilage and gums can aid in the chelation of essential micronutrients like iron, magnesium, and calcium. In many soils, these nutrients are present in insoluble forms, making them difficult for plants to absorb. The chelation process, facilitated by mucilage and gums, binds these nutrients in a form that is more readily available to plant roots, leading to improved nutrient uptake.

Research has shown that mucilage can enhance the availability of iron and magnesium in soils by up to 30%, which significantly improves the health and productivity of crops that rely on these nutrients for growth. By making these nutrients more available, mucilage contributes to healthier plants and higher crop yields [29].

9.6.5.6 Reduction of Fertilizer Dependency

By improving nutrient cycling and the availability of essential nutrients, mucilage and gums can reduce the need for synthetic fertilizers. This not only helps lower input costs for farmers but also minimizes the environmental impact associated with the excessive use of chemical fertilizers. The reduction in fertilizer use decreases the risk of nutrient runoff, which is a leading cause of soil degradation and water pollution.

Farmers can achieve significant cost savings by incorporating mucilage and gums into their agricultural practices. In some cases, the use of mucilage can reduce fertilizer costs by up to 40%, making it an attractive and cost-effective alternative to traditional fertilization methods [36].

9.6.6 Enhancing Soil Microbial Health

Healthy soils are teeming with microorganisms that play a critical role in nutrient cycling, disease suppression, and organic matter decomposition. Mucilage and gums support soil microbial health by providing a carbon source for beneficial microbes, creating a favorable environment for the growth of diverse microbial communities [28].

9.6.6.1 Microbial Habitat Creation

Mucilage and gums serve as an ideal carbon source for beneficial soil microbes, such as nitrogen-fixing bacteria, mycorrhizal fungi, and decomposers. These microorganisms help break down organic matter, making nutrients available to plants in a form that they can absorb. By fostering a healthy microbial community, mucilage and gums enhance nutrient cycling and improve soil fertility.

9.6.6.2 Increased Microbial Diversity

Soil microbial diversity is essential for maintaining soil health and ecosystem function. Mucilage and gums support microbial diversity by providing a range of nutrients and organic materials that benefit different types of soil organisms. A diverse microbial community enhances soil fertility by carrying out various essential functions, such as decomposing organic matter, fixing nitrogen, and suppressing harmful pathogens.

Studies have shown that the addition of mucilage to the soil can increase microbial diversity by up to 25%, leading to improved soil health and enhanced nutrient cycling. This increased diversity also helps maintain the balance of beneficial microorganisms, preventing the dominance of harmful pathogens and reducing the need for chemical pesticides [19].

9.6.6.3 Soil Enzyme Activity

Soil enzymes are produced by microorganisms and play a crucial role in the breakdown of organic matter and the release of nutrients. Mucilage and gums stimulate microbial growth, leading to increased enzyme production in the soil. This enhanced enzyme activity improves nutrient cycling and supports plant growth, ultimately boosting crop yield.

The increased enzyme activity associated with mucilage application can result in a 15–20% increase in nutrient availability, further contributing to the overall health and productivity of crops. In particular, enzyme activity helps release essential nutrients like nitrogen and phosphorus, which are key for optimal plant growth [25].

9.6.6.4 Increasing Crop Growth and Yield

One of the primary goals of any agricultural practice is to increase crop yield while maintaining environmental sustainability. Mucilage and gums contribute to higher yields by improving soil structure, enhancing nutrient availability, supporting microbial health, and promoting better root growth [27].

9.6.6.5 Better Root Development

The improved soil structure and aggregation facilitated by mucilage and gums create a more favorable environment for root growth. Plants with healthier, deeper roots are better able to access water and nutrients from a larger soil volume. This increased root depth supports more robust plant growth and enhances crop productivity.

By promoting root expansion, mucilage and gums ensure that plants can better withstand environmental stressors such as drought, nutrient deficiency, and soil compaction. As a result, crops are able to achieve higher growth rates and improved yield potential [30].

9.6.6.6 Improved Plant Stress Resistance

Mucilage's water-retaining properties help reduce water stress during periods of drought. By maintaining consistent moisture levels in the soil, mucilage ensures that plants experience less water stress, even during dry spells. This is especially important for crops grown in arid and semiarid regions, where drought can significantly reduce yield.

In addition to drought tolerance, mucilage and gums help improve plant disease resistance by promoting a healthy soil ecosystem. The increased microbial diversity in the soil helps suppress harmful pathogens and reduce the incidence of root diseases. This leads to healthier plants that are better able to withstand pest and disease pressure [31].

9.6.6.7 Increased Photosynthesis

With better nutrient uptake, stronger root systems, and enhanced resilience to environmental stress, plants are able to perform photosynthesis more effectively. Photosynthesis is the process by which plants convert sunlight into energy, and it is critical for growth and productivity. By supporting optimal photosynthesis, mucilage and gums help plants produce more energy, leading to stronger, healthier crops and ultimately higher yields [37].

9.6.6.8 Reducing Soil Erosion

Soil erosion is a major concern in many agricultural systems, especially in regions prone to heavy rainfall or high winds. Mucilage and gums help reduce soil erosion by binding soil particles together and promoting soil aggregation [30].

9.6.6.8.1 Erosion Control

The sticky properties of mucilage and gums help bind soil particles together, creating stable aggregates that are less prone to erosion. These aggregates help protect the soil from wind and water erosion, which can lead to the loss of valuable topsoil and nutrients.

In areas where erosion is a concern, the application of mucilage and gums can reduce soil erosion by up to 30%, preserving topsoil and preventing the loss of nutrients that are essential for crop growth [29].

9.6.6.8.2 Sustaining Topsoil

Topsoil is the most fertile layer of soil, containing essential nutrients and organic matter necessary for plant growth. By preventing erosion, mucilage and gums help preserve the integrity of topsoil, ensuring that soils remain productive for future generations. Sustainable management of topsoil is key to long-term agricultural productivity and environmental health [26].

9.6.7 Long-Term Soil Health and Sustainability

Soil health is critical for sustainable agricultural practices. Mucilage and gums play a vital role in enhancing long-term soil health by improving soil organic matter content, preventing degradation, and supporting soil microbial communities.

9.6.7.1 Soil Organic Matter Increase

Mucilage and gums contribute to the buildup of organic matter in the soil, which is essential for soil fertility. Organic matter decomposes over time to form humus, a stable component of soil that improves its structure, water-holding capacity, and nutrient availability. Increased organic matter also promotes beneficial microbial activity and supports long-term soil health.

The addition of mucilage and gums can increase soil organic matter by up to 15%, contributing to the overall sustainability and fertility of the soil.

9.6.7.2 Reduced Soil Degradation

By improving soil structure, supporting microbial health, and preventing erosion, mucilage and gums help reduce the risk of soil degradation. Healthy soils are less prone to compaction, erosion, and nutrient depletion, ensuring that agricultural land remains productive and viable over the long term.

9.6.7.3 Enhancing Soil pH Balance

Mucilage and gums can help buffer soil pH, preventing it from becoming too acidic or alkaline. This buffering action ensures that the soil remains within the optimal pH range for plant growth and supports microbial life in the soil.

9.6.8 Environmentally Friendly and Cost-Effective

Mucilage and gums offer an environmentally friendly alternative to synthetic fertilizers and chemicals, reducing the ecological footprint of agricultural operations. By enhancing soil fertility, reducing water usage, and supporting microbial health, mucilage and gums contribute to more sustainable farming practices.

9.6.8.1 Reduction in Chemical Inputs

By improving nutrient availability, supporting microbial activity, and reducing erosion, mucilage and gums reduce the need for synthetic fertilizers and pesticides. This not only lowers costs for farmers but also minimizes the environmental impact associated with the use of chemical inputs.

The use of mucilage and gums can reduce fertilizer dependence by up to 40%, making them a cost-effective and environmentally friendly alternative to traditional agricultural inputs.

9.6.8.2 Sustainability

Mucilage and gums contribute to the sustainability of farming systems by supporting natural soil processes and reducing the need for synthetic inputs. By promoting healthy soils, enhancing crop productivity, and reducing environmental impact, mucilage and gums play a critical role in advancing sustainable agriculture [15, 21].

References

[1]   Shiam, M. A. H., Islam, M. S., Ahmad, I., & Haque, S. S. (2025). A review of plant-derived gums and mucilages: Structural chemistry, film forming properties and application. *Journal of Plastic Film & Sheeting*, *41*(2), 195–237.

[2]   Campos, E. V. R., de Oliveira, J. L., Fraceto, L. F., & Singh, B. (2015). Polysaccharides as safer release systems for agrochemicals. *Agronomy for sustainable development*, *35*(1), 47–66.

[3]   Elhassan, G. A., Abdelgani, M. E., Osman, A. G., Mohamed, S. S., & Abdelgadir, B. S. (2010). Potential production and application of biofertilizers in Sudan. *Pakistan Journal of Nutrition*, *9*(9), 926–934.

[4]   Chittora, D., Meena, M., Barupal, T., Swapnil, P., & Sharma, K. (2020). Cyanobacteria as a source of biofertilizers for sustainable agriculture. *Biochemistry and biophysics reports*, *22*, 100737.

[5]   El-Sawah, A. M., El-Keblawy, A., Ali, D. F. I., Ibrahim, H. M., El-Sheikh, M. A., Sharma, A., . . . & Sheteiwy, M. S. (2021). Arbuscular mycorrhizal fungi and plant growth-promoting rhizobacteria enhance soil key enzymes, plant growth, seed yield, and qualitative attributes of guar. *Agriculture*, *11*(3), 194.

[6]   Singh, S. K., Wu, X., Shao, C., & Zhang, H. (2022). Microbial enhancement of plant nutrient acquisition. *Stress Biology*, *2*(1), 3.

[7]   Chittora, D., Meena, M., Barupal, T., Swapnil, P., & Sharma, K. (2020). Cyanobacteria as a source of biofertilizers for sustainable agriculture. *Biochemistry and biophysics reports*, *22*, 100737.

[8]   Hassan, H., Nassar, D., & Abou-Bakr, M. H. A. (2006). Effect of mineral and biofertilizers on growth, yield components, chemical constituents and anatomical structure of moghat plant (Glossostemon bruguieri Desf.) grown under reclaimed soil conditions. *Journal of Plant Production*, *31*(3), 1433–1455.

[9]   Elhassan, G. A., Abdelgani, M. E., Osman, A. G., Mohamed, S. S., & Abdelgadir, B. S. (2010). Potential production and application of biofertilizers in Sudan. *Pakistan Journal of Nutrition*, *9*(9), 926–934.

[10]   Tosif, M. M., Najda, A., Bains, A., Kaushik, R., Dhull, S. B., Chawla, P., & Walasek-Janusz, M. (2021). A comprehensive review on plant-derived mucilage: characterization, functional properties, applications, and its utilization for nanocarrier fabrication. *Polymers*, *13*(7), 1066.

[11]   Tsai, A. Y. L., McGee, R., Dean, G. H., Haughn, G. W., & Sawa, S. (2021). Seed mucilage: biological functions and potential applications in biotechnology. *Plant and Cell Physiology*, *62*(12), 1847–1857.

[12]   Dybka-Stępień, K., Otlewska, A., Góźdź, P., & Piotrowska, M. (2021). The renaissance of plant mucilage in health promotion and industrial applications: A review. *Nutrients*, *13*(10), 3354.

[13]   Kučka, M., Ražná, K., Harenčár, Ľ., & Kolarovičová, T. (2022). Plant seed mucilage – Great potential for sticky matter. *Nutraceuticals*, *2*(4), 253–269.

[14]   Ritzoulis, C. (2017). Mucilage formation in food: A review on the example of okra. *International Journal of Food Science and Technology*, *52*(1), 59–67.

[15]   Di Marsico, A., Scrano, L., Labella, R., Lanzotti, V., Rossi, R., Cox, L., . . . & Amato, M. (2018). Mucilage from fruits/seeds of chia (Salvia hispanica L.) improves soil aggregate stability. *Plant and Soil*, *425*(1), 57–69.

[16]   Beigi, S., Azizi, M., & Iriti, M. (2020). Application of super absorbent polymer and plant mucilage improved essential oil quantity and quality of Ocimum basilicum var. Keshkeni Luvelou. *Molecules*, *25*(11), 2503.

[17]   Jones, B. O., John, O. O., Luke, C., Ochieng, A., & Bassey, B. J. (2016). Application of mucilage from Dicerocaryum eriocarpum plant as biosorption medium in the removal of selected heavy metal ions. *Journal of environmental management*, *177*, 365–372.

[18]   Moglia, M., Cook, S., & Tapsuwan, S. (2018). Promoting water conservation: where to from here?. *Water*, *10*(11), 1510.

[19]   Razzaghi, F., Obour, P. B., & Arthur, E. (2020). Does biochar improve soil water retention? A systematic review and meta-analysis. *Geoderma*, *361*, 114055.

[20]   Boutraa, T. (2010). Improvement of water use efficiency in irrigated agriculture: a review. *Journal of Agronomy*, *9*(1), 1–8.

[21]   Ray, R. L., Ampim, P. A., & Gao, M. (2020). Crop protection under drought stress. *Crop protection under changing climate*, 145–170.

[22]   Xiong, M., Sun, R., & Chen, L. (2018). Effects of soil conservation techniques on water erosion control: A global analysis. *Science of the Total Environment*, *645*, 753–760.

[23] Jiang, Y., Hou, L., Ding, F., & Whalen, J. K. (2023). Root mucilage: Chemistry and functions in soil. *Encyclopedia of Soils in the Environment, 1*, 332–42.

[24] Ebrahimi, A., Moaveni, P., & Farahani, H. A. (2010). Effects of planting dates and compost on mucilage variations in borage (Borago officinalis L.) under different chemical fertilization systems. *International Journal for Biotechnology and Molecular Biology Research, 1*(5), 58–61.

[25] Nazari, M. (2021). Plant mucilage components and their functions in the rhizosphere. *Rhizosphere, 18*, 100344.

[26] Di Marsico, A., Scrano, L., Amato, M., Gàmiz, B., Real, M., & Cox, L. (2018). Influence of mucilage from seeds of chia (Salvia hispanica L.) as an amendment, on the sorption-desorption of herbicides in agricultural soils. *Science of the Total Environment, 625*, 531–538.

[27] Nazari, M., Bilyera, N., Banfield, C. C., Mason-Jones, K., Zarebanadkouki, M., Munene, R., & Dippold, M. A. (2023). Soil, climate, and variety impact on quantity and quality of maize root mucilage exudation. *Plant and Soil, 482*(1), 25–38.

[28] Poshkoohi, A. F., Mohammadi, M. H., Zarebanadkouki, M., & Etesami, H. (2025). Mucilage increases soil resistance to penetration after compaction. *Rhizosphere, 33*, 101014.

[29] Riseh, R. S., Tamanadar, E., Pour, M. M., & Thakur, V. K. (2022). Novel approaches for encapsulation of plant probiotic bacteria with sustainable polymer gums: application in the management of pests and diseases. *Advances in polymer technology, 2022*(1), 4419409.

[30] Ahmed, M. A., Sanaullah, M., Blagodatskaya, E., Mason-Jones, K., Jawad, H., Kuzyakov, Y., & Dippold, M. A. (2018). Soil microorganisms exhibit enzymatic and priming response to root mucilage under drought. *Soil Biology and Biochemistry, 116*, 410–418.

[31] Shiam, M. A. H., Islam, M. S., Ahmad, I., & Haque, S. S. (2025). A review of plant-derived gums and mucilages: Structural chemistry, film forming properties and application. *Journal of Plastic Film & Sheeting, 41*(2), 195–237.

[32] Haruna, S., Aliyu, B. S., & Bala, A. (2016). Plant gum exudates (Karau) and mucilages, their biological sources, properties, uses and potential applications: A review. *Bayero Journal of Pure and Applied Sciences, 9*(2), 159–165.

[33] Darrell, N. (2020). *Conversations with plants: the path back to nature*. Aeon Books.

[34] Engelbrecht, M., Bochet, E., & García-Fayos, P. (2014). Mucilage secretion: an adaptive mechanism to reduce seed removal by soil erosion?. *Biological Journal of the Linnean Society, 111*(2), 241–251.

[35] Di Marsico, A., Scrano, L., Labella, R., Lanzotti, V., Rossi, R., Cox, L., . . . & Amato, M. (2018). Mucilage from fruits/seeds of chia (Salvia hispanica L.) improves soil aggregate stability. *Plant and Soil, 425*(1), 57–69.

[36] Dingley, C., Cass, P., Adhikari, B., Bendre, P., Mantri, N., & Daver, F. (2024). Psyllium husk mucilage as a novel seed encapsulant for agriculture and reforestation. *Journal of Sustainable Agriculture and Environment, 3*(3), e70004.

[37] Siam, A. M., & Abdullah, N. A. (2022). PHOTOCHEMICAL EFFICIENCY AND GROWTH OF TWO PROVENANCES OF ACACIA SENEGAL (GUM ARABIC (TREE SEEDLINGS UNDER DRYING SOIL. *Journal of Environmental Science, 51*(4), 1–30.

10 Cosmetics and Personal Care Products

10.1 Introduction

Mucilages and gums have gained significant attention in the cosmetics and personal care industry due to their multifunctional properties and natural origin [1]. These plant-derived polysaccharides are increasingly being incorporated into skincare and haircare formulations as effective, biodegradable alternatives to synthetic ingredients. In skincare, mucilages and gums act as excellent hydrating agents by forming a moisture-retentive film on the skin's surface. Their film-forming and water-binding capabilities help maintain skin hydration, improve texture, and provide a soothing effect, making them ideal for use in moisturizers, lotions, and serums [2]. They also enhance the stability and consistency of emulsions, contributing to smoother, creamier textures in cosmetic products.

In haircare formulations, mucilages and gums offer conditioning benefits by coating hair strands, reducing frizz, and improving manageability. Their natural emulsification properties assist in stabilizing oil and water mixtures in shampoos and conditioners, enhancing product performance. Additionally, their gentle and nonirritant nature makes them suitable for sensitive skin and scalp formulations. Mucilages and gums in personal care products contribute to the growing demand for clean-label, eco-conscious beauty items by replacing potentially harmful synthetic thickeners, emulsifiers, and stabilizers [2, 3].

These natural polymers also support the delivery of active ingredients in both skin and hair applications, enabling sustained release and improved absorption. As consumers increasingly seek plant-based and sustainable options, mucilages and gums offer an effective solution that meets both performance and environmental standards. Their multifunctional roles in hydration, film formation, and emulsification not only enhance product functionality but also promote healthier skin and hair [4]. Thus, mucilages and gums are proving to be essential components in the development of innovative, safe, and eco-friendly cosmetics and personal care products.

10.2 Mucilage and Gums in Skin Care and Hair Care Formulations

Mucilage is a viscous, gel-like substance that plants produce to retain water and protect themselves from environmental stressors. This gel-like consistency comes from the complex carbohydrates such as polysaccharides and glycoproteins, which make up mucilage. Plants that are typically high in mucilage include aloe vera, flaxseed, marshmallow root, and okra. The unique ability of mucilage to absorb and retain large amounts of water gives it a prominent place in hydrating products [4].

https://doi.org/10.1515/9783111673509-010

The primary role of mucilage in nature is to help plants conserve water, making it an ideal ingredient for hydrating and moisturizing skin and hair. When applied to the skin or hair, mucilage creates a barrier that locks in moisture, providing long-lasting hydration without making the skin or hair greasy or weighed down. It also has natural soothing properties, which is why mucilage is widely used in products for sensitive skin or those suffering from conditions like eczema, psoriasis, and sunburns [5].

10.2.1 Benefits of Mucilage in Skin Care [6]

Mucilage's ability to retain water makes it an excellent choice for maintaining skin hydration. When applied topically, mucilage works by attracting moisture from the environment and locking it into the skin, preventing dehydration. This ability to bind moisture and hold it on the skin's surface helps keep the skin soft, plump, and youthful. Some of the most notable benefits of mucilage in skin care include:

- **Moisturizing:** Mucilage is highly hydrophilic, meaning it attracts and holds water. This makes it an effective moisturizer, particularly in dry climates or for those with dry skin. The water-binding ability helps prevent moisture loss, keeping the skin hydrated for longer periods of time.
- **Soothing and Anti-inflammatory:** Many mucilage-rich plants, such as aloe vera, contain natural anti-inflammatory compounds that help reduce redness and irritation. This makes mucilage a great option for calming inflamed or irritated skin, particularly for conditions like sunburns, eczema, or psoriasis. Its gentle nature makes it ideal for sensitive skin.
- **Healing:** Mucilage's soothing properties also contribute to its healing abilities. For example, aloe vera has been used for centuries as a natural remedy for wounds, burns, and cuts due to its regenerative effects. It accelerates the healing process by promoting cell turnover and improving skin elasticity.
- **Film-Forming:** Mucilage creates a breathable, protective layer on the skin's surface. This thin film acts as a barrier that shields the skin from environmental stressors such as pollutants, harsh weather, and UV rays, while still allowing the skin to breathe and retain moisture.

10.2.2 Benefits of Mucilage in Hair Care

Just as mucilage is beneficial for skin, it also provides numerous advantages for hair care. When used in hair products, mucilage helps hydrate and condition the hair, especially for individuals with dry, frizzy, or curly hair types. Mucilage helps maintain moisture in the hair fibers, improving their elasticity and reducing the likelihood of breakage [7]. Some of the key benefits of mucilage in hair care include:

- **Hydration:** Like its effects on the skin, mucilage hydrates hair by attracting moisture and locking it into the hair shaft. This is particularly helpful for people with dry, brittle, or coarse hair that is prone to frizz and breakage.
- **Frizz Control:** Mucilage helps control frizz by forming a protective barrier around the hair, which prevents moisture from escaping and the hair from becoming dehydrated. This makes it a key ingredient in anti-frizz hair care products, especially for curly or wavy hair types.
- **Scalp Health:** Mucilage can also benefit the scalp by soothing irritation, reducing inflammation, and balancing moisture levels. A healthy scalp is essential for promoting hair growth, and mucilage helps to create an ideal environment for hair follicles to thrive.
- **Light Styling and Hold:** Mucilage-rich ingredients like aloe vera and flaxseed are often used in styling products, such as hair gels, to provide light hold without the sticky or crunchy texture that can come with synthetic styling agents. Mucilage helps define curls, tame flyaways, and add texture while maintaining hydration [8].

10.2.3 Examples of Mucilage in Skin and Hair Care Products

Several plants rich in mucilage are widely used in cosmetic formulations for both skin and hair care. Aloe vera is perhaps the most well-known, with its cooling and hydrating effects making it a popular ingredient in moisturizers, shampoos, and conditioners. Aloe vera's mucilage helps calm and hydrate the skin and hair, promoting healing and improving elasticity.

Flaxseed is another mucilage-rich plant commonly found in hair care products like gels and conditioners. Flaxseed mucilage is highly hydrating and helps to nourish the hair, reducing frizz and improving texture. Marshmallow root, with its soothing properties, is often included in hair conditioners and skin lotions to improve hydration and manageability. Okra, though less common, is also known for its high mucilage content and is sometimes used in hair masks and treatments for its detangling and moisturizing benefits [5, 7].

10.2.4 What Are Gums in Skin and Hair Care?

Gums are natural exudates (resins) produced by various plants, typically trees, as a form of defense. These gums, composed of complex polysaccharides, share many properties with mucilage but tend to be more solid and less viscous. Common gums used in the cosmetic industry include xanthan gum, guar gum, acacia gum, and gum arabic. Gums are valued primarily for their ability to stabilize emulsions, thicken products, and provide moisture retention [9].

Gums are a key ingredient in emulsifying systems that blend oil and water in skin and hair care formulations. Without emulsifiers like gums, these mixtures would separate, making the products less effective and aesthetically pleasing. Gums help improve the texture, consistency, and performance of skin and hair care formulations, enhancing the overall user experience [10].

10.2.5 Benefits of Gums in Skin Care [11]

Gums offer several benefits when incorporated into skin care products, with the most notable being their ability to stabilize emulsions and improve product texture. Gums are often used in products like moisturizers, serums, and masks to enhance their performance and provide a more luxurious feel. Some of the main benefits of gums in skin care include:

- **Emulsion Stabilization:** Gums help prevent the separation of water and oil in emulsions, which are commonly found in creams, lotions, and serums. By stabilizing the mixture, gums ensure a smooth, consistent texture throughout the product's shelf life.
- **Thickening Agents:** Gums are often used to thicken formulations, adding viscosity to the product without making it greasy or heavy. This is particularly useful in products like gels and creams, where the texture is a crucial aspect of the product's performance.
- **Moisturizing:** Like mucilage, gums help retain moisture by forming a barrier that prevents water loss from the skin. This barrier also helps protect sensitive skin from environmental stressors, such as harsh weather or pollutants.
- **Film-Forming:** Gums can create a thin, protective layer on the skin, which helps shield it from external irritants while maintaining hydration. This makes gums useful in products like sunscreens or treatments for sensitive skin.

10.2.6 Benefits of Gums in Hair Care [12]

In hair care formulations, gums provide conditioning effects, texture enhancement, and moisture retention. Their ability to create a smooth, hydrated barrier around the hair strands makes them ideal for use in shampoos, conditioners, and styling products. Some of the notable benefits of gums in hair care include:

- **Conditioning:** Gums help coat the hair, providing a smooth, soft finish that makes the hair more manageable. This conditioning effect is especially beneficial for dry, damaged, or coarse hair types.
- **Volume and Thickness:** Gums like xanthan gum can be used to add body and thickness to hair products without the need for silicones. This helps create fuller, more voluminous hairstyles.

– **Frizz Control:** Similar to mucilage, gums can help prevent frizz by sealing moisture in the hair shaft and creating a smooth, sleek appearance. Gums help maintain hydration in the hair while taming flyaways.

10.2.7 Examples of Gums in Skin and Hair Care Products

Xanthan gum is one of the most commonly used gums in both skin and hair care products. It is valued for its ability to thicken and stabilize emulsions, making it a popular choice for moisturizers, face masks, and shampoos. Guar gum, another popular gum, is often found in hair conditioners, shampoos, and lotions for its moisturizing and thickening effects. Acacia gum, also known as gum arabic, is frequently used in emulsions to create smooth textures and improve product stability [13].

10.3 Combining Mucilage and Gums in Formulations

When mucilage and gums are used together in cosmetic formulations, they offer a synergistic effect that enhances the overall performance of the product. Mucilage provides intense hydration and soothing benefits, while gums improve product stability, texture, and overall feel. This combination can be particularly effective in products like moisturizers, hair conditioners, and styling gels, where both hydration and smooth texture are key [11, 13].

For example, a moisturizer may contain aloe vera mucilage for its hydrating and healing properties, while xanthan gum could be used to thicken the formula and create a pleasant, silky texture. Similarly, a hair gel may combine flaxseed mucilage with guar gum to provide both hydration and hold, offering a product that nourishes the hair while providing structure and control [14].

10.3.1 Applications of Mucilage and Gums in Cosmetics [1, 5, 11]

The versatility of mucilage and gums makes them suitable for a wide range of cosmetic applications, from moisturizing products to hair styling agents. These natural ingredients not only improve product performance but also contribute to the growing trend of clean, sustainable beauty. By using plant-derived substances like mucilage and gums, cosmetics manufacturers can create products that are both effective and eco-friendly:

– **Skin Care Products:** Mucilage and gums are commonly used in moisturizers, face masks, and serums to hydrate, soothe, and protect the skin. Aloe vera gel, flaxseed mucilage, and xanthan gum are often found in these formulations for their ability to improve skin texture and retain moisture.

- **Hair Care Products:** In shampoos, conditioners, and styling products, mucilage and gums provide conditioning, frizz control, and moisture retention. Aloe vera, flaxseed mucilage, and guar gum are commonly used in these formulations to improve hair health, manageability, and appearance.
- **Scalp Treatments:** Mucilage and gums are also beneficial in scalp treatments, where their soothing and moisturizing properties can help reduce irritation, dryness, and dandruff. Aloe vera mucilage is often included in these formulations to promote scalp health and encourage healthy hair growth.

10.4 Hydration, Film-Forming, and Emulsification Properties [7–11]

Hydration is one of the most significant benefits of mucilage and gums in skin and hair care formulations. The ability to attract and retain water is essential for maintaining moisture balance, preventing dryness, and ensuring that the skin and hair remain soft, supple, and healthy looking. Hydration influences everything, from the texture and elasticity of the skin to the strength and appearance of the hair.

10.4.1 How Hydration Works

Both mucilage and gums are **hydrophilic** (water-attracting), meaning they have the ability to draw moisture from the surrounding environment and lock it into the skin or hair. The presence of these ingredients in formulations helps ensure that the skin remains moisturized throughout the day and the hair stays hydrated without losing its natural moisture content. The process is not limited to just the initial absorption of water but extends to moisture retention, which keeps the skin and hair hydrated over time.

10.4.2 Moisture Retention

Once applied to the skin or hair, mucilage and gums help retain water in the outermost layers. For skin, this means that the water is maintained within the **stratum corneum**, the outermost layer of skin cells. In the hair, the mucilage and gums lock moisture within the hair shaft, preventing water loss that could lead to dryness or damage.

10.4.3 Benefits in Skin Care [5, 9]

- **Improved Skin Elasticity:** Well-hydrated skin maintains its natural elasticity, which is crucial for reducing the appearance of fine lines and wrinkles. Hydration helps support collagen and elastin fibers in the dermis, making the skin more resilient to environmental stress.
- **Softness and Smoothness:** Proper hydration is key to achieving soft, plump, and smooth skin. Skin that is well-moisturized has a more youthful and radiant appearance, as it is less prone to dryness, flakiness, and irritation.
- **Barrier Repair:** Hydrated skin strengthens its natural barrier function, which is essential for protection from environmental stressors such as pollutants, UV rays, and harsh weather conditions. A strong skin barrier prevents excessive moisture loss, maintaining the skin's overall health and appearance.

10.4.4 Benefits in Hair Care [5, 8]

Shiny, Glossy Hair: Hydrated hair tends to reflect light more effectively, making it appear shinier and more vibrant. Water acts as a natural light-reflecting agent, enhancing the glossy appearance of hair:
- **Prevention of Breakage:** When hair is properly hydrated, it is less likely to become brittle and prone to breakage. Moisturizing ingredients help improve hair elasticity, reducing the chances of hair snapping during styling or detangling.
- **Manageability:** Well-hydrated hair is easier to detangle and style. Moisture helps to smooth the cuticle layer of the hair, making it more manageable and less frizzy, thus improving overall styling efficiency.

10.4.5 Examples of Mucilage and Gums for Hydration

- **Aloe Vera Mucilage:** Aloe vera is one of the most popular mucilage-rich plants known for its highly hydrating properties. Aloe vera gel can be used in both skin and hair care products to deeply hydrate and nourish [15].
- **Flaxseed Gel:** Flaxseed mucilage is a highly effective hydrating agent used in various hair care products, such as hair gels and conditioners. It helps lock in moisture, enhancing hair texture and preventing frizz [16].
- **Marshmallow Root Mucilage:** Known for its soothing and hydrating properties, marshmallow root is often found in skin care products to promote hydration and calm irritated skin. It can also be used in hair care products to moisturize the scalp and hair [17].

10.5 Film-Forming Properties

Film-forming agents are crucial for creating protective layers on the skin and hair. These agents form thin, flexible, and breathable films that serve several important functions: protecting moisture, enhancing product performance, and offering a smooth finish. The ability of mucilage and gums to form such films makes them highly beneficial in both skin and hair care formulations [8].

10.5.1 How Film-Forming Works

Many mucilage and gums are **water-soluble**, meaning they can create films that are breathable but still effective at locking in moisture. When applied, they form a thin layer over the skin or hair that shields from environmental stressors while allowing the skin to breathe. This barrier helps to prevent the evaporation of moisture, contributing to long-lasting hydration [12].

10.5.2 Protective Layer

The films created by these ingredients serve as a physical barrier to pollutants, wind, UV radiation, and other harmful factors that can damage the skin or hair. The thin layer of gum or mucilage prevents these harmful substances from penetrating the skin or hair, protecting it from external environmental factors [11].

10.5.3 Cosmetic Effect

Apart from protection, the film-forming properties of mucilage and gums also provide a cosmetic effect. The thin layer enhances the appearance of both skin and hair, making them appear smoother and look more radiant. For instance, on the skin, it provides a silky and nongreasy finish, while in hair care, it contributes to a smooth, sleek, and frizz-free look [2].

10.5.4 Benefits in Skin Care [2, 3]

- **Environmental Protection:** The film forms a barrier that shields the skin from pollutants, environmental irritants, and UV rays while still allowing the skin to breathe. This is particularly beneficial for those with sensitive skin who are prone to irritation or environmental damage.

– **Long-Lasting Hydration:** By preventing moisture from evaporating, the film ensures that the skin remains hydrated for a longer time. The protective barrier helps retain water in the skin, ensuring that hydration levels stay consistent throughout the day.
– **Improved Texture:** The film also enhances the texture of creams, lotions, and serums, contributing to a smooth and pleasant application. The film-forming property of mucilages and gums can make the product feel luxurious and nongreasy, without leaving a heavy or sticky residue.

10.5.5 Benefits in Hair Care [13, 14]

– **Frizz Control:** One of the most notable benefits of the film-forming property is its ability to control frizz. The film smooths the hair's surface, reducing the impact of humidity and making the hair look sleek and polished.
– **Shine:** The film enhances the natural shine of hair by reflecting light, creating a glossy and healthy appearance. Well-moisturized hair looks shinier because the cuticles are laid flat, creating a smooth surface for light reflection.
– **Protection:** The film also provides protection against environmental damage caused by heat styling, pollution, or excessive sun exposure. By sealing in moisture, it helps to prevent hair from becoming dry, brittle, and prone to damage.

10.5.6 Examples of Mucilages and Gums for Film-Forming:

– **Gum Arabic:** Known for its excellent film-forming properties, gum arabic is used in a variety of skin care formulations to create a protective barrier. It helps in shielding the skin from pollutants and keeping moisture locked in [18].
– **Xanthan Gum:** While primarily used as a thickener, xanthan gum also creates a breathable film on the skin that helps retain moisture and improve hydration [19].
– **Aloe Vera Gel:** Aloe vera is not only hydrating but also forms a protective layer on the skin and hair, making it ideal for both soothing and moisturizing properties [15].

10.6 Emulsification Properties

Emulsification is the process of mixing two liquids that do not naturally combine, such as oil and water. In cosmetic formulations, emulsifiers are used to stabilize mixtures of water-based and oil-based ingredients, ensuring that the product maintains a smooth, homogeneous consistency. Both mucilages and gums are excellent emulsifiers, as they help to bind water and oil together, preventing separation and ensuring that the product performs consistently throughout its shelf life [20].

10.6.1 How Emulsification Works

Emulsifiers like mucilages and gums contain molecules with both **hydrophilic (water-attracting)** and **lipophilic (oil-attracting)** properties. These molecules create a stable emulsion by binding the water and oil molecules together. By doing so, they ensure that the formulation remains smooth and stable. Many emulsifiers also have the ability to **control viscosity**, thickening the product for a more luxurious feel [20].

10.6.2 Prevention of Phase Separation

Emulsifiers prevent the separation of the oil and water phases in products like lotions, creams, shampoos, and conditioners. Without emulsifiers, oil and water would naturally separate, rendering the product ineffective and aesthetically unappealing. Gums and mucilages ensure that the emulsion remains consistent, providing an even application and long-lasting stability [20].

10.6.3 Benefits in Skin Care [21]

– **Improved Product Texture:** Emulsifiers contribute to the smooth, creamy texture of skin care products like creams, lotions, and serums. The consistency provided by emulsifiers enhances the ease of application and absorption.
– **Even Distribution of Ingredients:** By stabilizing the emulsion, emulsifiers ensure that active ingredients like vitamins, oils, and botanical extracts are evenly distributed throughout the formulation, enhancing the effectiveness of the product.
– **Stable Formulations:** The use of emulsifiers extends the shelf life of skin care products. By preventing separation and maintaining consistency, emulsifiers ensure that the product remains effective and high-quality over time.

10.6.4 Benefits in Hair Care [20, 21]

– **Shampoo and Conditioner Formulations:** Emulsifiers ensure that the water, oils, and surfactants in shampoos and conditioners remain blended. This ensures that the product applies evenly and functions as intended, providing consistent cleansing and conditioning.

- **Hair Treatments:** In treatments like hair masks or serums, emulsifiers help combine water-based and oil-based ingredients. This ensures that the active ingredients penetrate the hair shaft, providing nourishment and improving hair health.
- **Detangling:** Emulsifying agents help coat the hair strands, making them smoother and less prone to tangling. This results in easier combing and brushing, reducing breakage and promoting smoother, more manageable hair.

10.6.5 Examples of Mucilages and Gums for Emulsification

- **Xanthan Gum:** A widely used emulsifier, xanthan gum stabilizes emulsions and thickens formulations, ensuring a smooth texture in skin and hair care products [19].
- **Guar Gum:** Guar gum is another natural emulsifier commonly found in hair care products. It improves the consistency of shampoos and conditioners while also providing moisturizing benefits [22].
- **Acacia Gum (Gum Arabic):** Known for its excellent emulsifying properties, gum arabic is used in a variety of cosmetic formulations, especially in high-end skin care products [23].

10.7 Natural Alternatives to Synthetic Ingredients in Personal Care

As the demand for natural products continues to rise, many people are seeking alternatives to synthetic chemicals in personal care products.

10.7.1 Natural Surfactants: Alternatives to Synthetic Cleansers [24–26]

Surfactants are key ingredients in personal care products, especially cleansers, shampoos, and body washes, as they reduce the surface tension of water, helping to remove dirt, oil, and impurities from the skin and hair. However, many synthetic surfactants like **sodium lauryl sulfate** and **sodium laureth sulfate** can be harsh on the skin, causing dryness or irritation, especially for sensitive skin types.

10.7.2 Natural Alternatives

- **Cocamidopropyl Betaine:** This is a mild surfactant derived from coconut oil. It has a lower irritancy potential than conventional surfactants and is often used in

shampoos, body washes, and facial cleansers. Its gentle cleansing properties make it suitable for sensitive skin.
- **Saponified Oils**: Saponification is a natural process that occurs when oils like **olive oil** or **coconut oil** are mixed with an alkali, creating soap. This soap gently cleanses the skin, without the harshness of synthetic surfactants.
- **Decyl Glucoside**: This is a nonionic surfactant derived from **corn glucose** and **coconut oil**. It is extremely gentle on the skin and is often found in baby products and formulas for sensitive skin. Decyl glucoside doesn't strip the skin of its natural oils, making it an excellent choice for daily cleansing.
- **Cocoyl Glutamate**: Derived from **coconut oil** and **glutamic acid** (an amino acid), this surfactant is often used in facial cleansers and shampoos. It is known for its mildness and ability to cleanse without disrupting the skin's natural moisture balance.

10.7.3 Natural Preservatives: Alternatives to Synthetic Preservatives [27–30]

Preservatives play a crucial role in personal care products, preventing microbial growth and extending shelf life. However, synthetic preservatives like **parabens** have been linked to potential health issues, including hormone disruption. Many consumers are now opting for natural preservatives that can offer similar antimicrobial protection without the harmful side effects.

10.7.4 Natural Alternatives

- **Vitamin E (Tocopherol)**: Vitamin E is a natural antioxidant that helps prevent the oxidation of oils in skincare products. By extending shelf life and protecting the product from degradation, it serves as an effective preservative.
- **Rosemary Extract**: This plant extract has powerful antimicrobial properties and is commonly used in natural formulations to prevent bacterial growth. Rosemary extract also has antioxidant benefits, which helps to protect both the product and the skin.
- **Grapefruit Seed Extract**: Known for its antibacterial and antifungal properties, grapefruit seed extract can serve as a natural preservative in cosmetics, preventing the growth of harmful microorganisms.
- **Radish Root Ferment Filtrate**: Derived from the fermentation of **radish root**, this preservative is becoming a popular alternative to parabens and other synthetic preservatives. It has been shown to have antimicrobial activity, making it an excellent natural preservative option.

10.8 Natural Moisturizers: Alternatives to Synthetic Emollients [31–33]

Emollients help moisturize and soften the skin. While synthetic emollients like **Dimethicone** are often used for their smooth texture and long-lasting effect, they can build up on the skin and block pores. Natural emollients are often preferred for their skin-soothing and nourishing properties.

10.8.1 Natural Alternatives

- **Shea Butter**: Shea butter is a rich, natural emollient that deeply nourishes the skin. It's rich in fatty acids and vitamins A, E, and F, which help to restore moisture and improve the skin's elasticity.
- **Cocoa Butter**: Derived from cocoa beans, cocoa butter is known for its ability to lock in moisture. It is often used in body butters and lotions due to its smooth texture and ability to hydrate the skin.
- **Aloe Vera Gel**: Known for its cooling and soothing properties, **aloe vera** is ideal for sensitive or irritated skin. It hydrates and provides relief for sunburns, dryness, and skin inflammation.
- **Jojoba Oil**: Structurally similar to the skin's natural oils (sebum), jojoba oil deeply moisturizes without clogging pores, making it suitable for all skin types, including acne-prone skin.
- **Coconut Oil**: Rich in fatty acids, **coconut oil** is a natural emollient that provides long-lasting moisture to both skin and hair. It helps protect the skin's natural barrier while nourishing and preventing dryness.

10.9 Natural Fragrances: Alternatives to Synthetic Fragrances [34–37]

Synthetic fragrances are often made from petrochemicals and can cause skin irritation and allergic reactions. These fragrances may also contain **phthalates**, which are chemicals linked to hormone disruption. For those looking for more natural options, essential oils and other plant-based fragrances are a better choice.

10.9.1 Natural Alternatives

- **Essential Oils**: Essential oils like **lavender, rose**, and **tea tree oil** provide not only natural fragrance but also therapeutic benefits. Lavender is known for its calming effects, while tea tree oil has antibacterial properties.

- **Floral Waters (Hydrosols)**: These are the distilled waters from flowers, like **rose water** or **chamomile water**. They offer delicate, natural fragrances and have additional benefits such as soothing and moisturizing the skin.
- **Citrus Extracts**: Essential oils and extracts from citrus fruits like **lemon**, **orange**, and **bergamot** offer refreshing, uplifting scents while providing antioxidant and anti-inflammatory benefits to the skin.
- **Spices and Herbs**: Ingredients like **cinnamon**, **ginger**, and **vanilla** offer warm, spicy fragrances that are safe for the skin and provide aromatic benefits without relying on synthetic chemicals.

10.10 Natural Colorants: Alternatives to Synthetic Dyes [38–41]

Synthetic dyes are often used to add color to cosmetic products but can be irritating to the skin, especially for those with sensitive skin types. Natural colorants are derived from plants, minerals, and other organic sources, offering a safer alternative.

10.10.1 Natural Alternatives

- **Beetroot Powder**: This powder provides a natural pink or red tint, often used in lip balms, blushes, and other cosmetics. It also contains antioxidants, which can be beneficial for the skin.
- **Spirulina Powder**: Spirulina is a blue-green algae that can provide a natural blue hue in cosmetics. It is rich in vitamins and minerals that nourish the skin.
- **Cocoa Powder**: Used to create a natural brown shade, cocoa powder is often found in bronzers, eyeshadows, and lip products.
- **Annatto Seed Extract**: Derived from the **achiote plant**, annatto provides a vibrant yellow or orange hue and has been used as a natural dye for centuries.
- **Turmeric**: Known for its anti-inflammatory properties, **turmeric** is often used to provide a golden or yellow tint to skincare and cosmetic products.

10.10.2 Natural Exfoliants: Alternatives to Synthetic Scrubs [42–44]

Exfoliants are used to remove dead skin cells and promote a smoother, healthier-looking complexion. Many synthetic exfoliants, like **microbeads**, are nonbiodegradable and contribute to environmental pollution. Natural exfoliants are not only gentle on the skin but also eco-friendly.

10.10.3 Natural Alternatives

–   **Sugar**: Sugar is a gentle, natural exfoliant that dissolves easily in water, making it both effective and eco-friendly. It helps to scrub away dead skin cells without causing irritation.
–   **Coffee Grounds**: Rich in antioxidants, **coffee grounds** are often used in body scrubs to exfoliate the skin. They also improve circulation and can help reduce the appearance of cellulite.
–   **Ground Almonds**: Ground almonds or **almond meal** offer a mild exfoliation that is gentle enough for sensitive skin. The natural oils in almonds also nourish the skin while exfoliating.
–   **Oatmeal**: Naturally soothing and anti-inflammatory, **oatmeal** serves as a gentle exfoliant while also calming the skin. It is often used in formulations for sensitive skin.

10.11 Natural Hair Conditioning Agents: Alternatives to Synthetic Silicones [45–48]

Silicones like **Dimethicone** are commonly used in hair care products to provide smoothness, shine, and manageability. However, silicones can leave a buildup on the hair, causing it to become dull and weighed down over time. Natural hair conditioning agents help nourish the hair without the risk of buildup.

10.11.1 Natural Alternatives

–   **Argan Oil**: Known for its high levels of **vitamin E** and essential fatty acids, **argan oil** is used to hydrate and condition hair. It helps restore shine and reduce frizz without weighing hair down.
–   **Coconut Oil**: **Coconut oil** is a popular natural conditioner that deeply nourishes the hair, helping to reduce frizz and improve shine without the use of silicones.
–   **Aloe Vera**: Aloe vera gel helps hydrate and condition hair without weighing it down. It also has soothing properties, making it ideal for sensitive scalps.
–   **Avocado Oil**: **Avocado oil** is rich in vitamins A, D, and E, as well as essential fatty acids. It helps nourish the hair, improving elasticity and leaving it shiny and manageable.

10.12 Natural Sunscreens: Alternatives to Chemical Sunscreens [49–52]

Chemical sunscreens like **oxybenzone** and **avobenzone** have been linked to potential hormonal effects and are harmful to marine life. For those seeking more natural sun protection, mineral-based sunscreens are a better option.

10.12.1 Natural Alternatives

- **Zinc Oxide**: **Zinc oxide** is a physical sunscreen that sits on the skin's surface and reflects UV rays. It is nonirritating and provides broad-spectrum protection without the chemical load.
- **Titanium Dioxide**: Another physical sunscreen, **titanium dioxide** provides protection against both UVA and UVB rays. It is often used in combination with zinc oxide for more effective sun protection.
- **Carrot Seed Oil**: While it does not provide high SPF protection, **carrot seed oil** offers natural UV protection and is often used in natural sunscreens as a supplementary ingredient.
- **Red Raspberry Seed Oil**: **Red raspberry seed oil** has a natural SPF of around 28–50, making it a popular option for natural sun protection when combined with other oils.

The growing trend toward natural alternatives in personal care products reflects a desire for safer, more eco-friendly options.

References

[1]   Uyor, U. O., Popoola, P. A., & Popoola, O. (2025). Advances in Mucilage Extraction Techniques and their Emerging Applications. *International Journal of Home Economics, Hospitality and Allied Research*, *4*(1), 39–63.

[2]   Martins, V. B., Da Silva Carvalho, J. G., PIETRO, B., GABRIELLI, A., ALVES DA CUNHA, M. A., KLEIN DAS NEVES, J. C., . . . & BUDZIAK PARABOCZ, C. R. (2021). Taro Mucilage: Extraction, Characterization, and Application in Cosmetic Formulations. *Journal of Cosmetic Science*, *72*(3).

[3]   There, U., Gour, N., Shrikant, S., Choudhary, V., & Kandasamy, R. (2023). Development of skincare formulations using flaxseed oil and mucilage. *Journal of Pharmacognosy and Phytochemistry*, *12*(2), 33–39.

[4]   Saraf, S. (2024). Investigating the Anti-Allergic, Bronchoprotective, and Antioxidant Activities of Psidium guajava Extract in Experimental Animals: A Comprehensive Study. *Journal of Internal Medicine and Pharmacology (JIMP)*, *1*(02), 33–43.

[5]   Uyor, U. O., Popoola, P. A., & Popoola, O. (2025). Advances in Mucilage Extraction Techniques and their Emerging Applications. *International Journal of Home Economics, Hospitality and Allied Research*, *4*(1), 39–63.

[6]   Moreno, W. Q., Proano-Molina, M., Proano-Molina, P., Mesa-Aguilar, J., Vera-Oyague, M., Caicedo-Álvarez, M., . . . & Ramirez-Gutierrez, C. (2023). Phytochemical and thickening properties of the mucilage of Malachra alceifolia jacq., in a shampoo formulation. *Afinidad. Journal of Chemical Engineering Theoretical and Applied Chemistry, 80*(598), 80–87.

[7]   Sitthithaworn, W., Khongkaw, M., Wiranidchapong, C., & Koobkokkruad, T. (2018). Mucilage powder from Litsea glutinosa leaves stimulates the growth of cultured human hair follicles. *Songklanakarin J. Sci. Technol, 40*, 1076–1080.

[8]   Sirajudheen, M. K., & Shijikumar, P. S. (2022). FORMULATION AND EVALUATION OF HAIR CONDITIONER CONTAINING HIBISCUS MUCILAGE AND VITAMIN E.

[9]   Gray, J. (2001). Hair care and hair care products. *Clinics in dermatology, 19*(2), 227–236.

[10]   Kapoor, V. P. (2005). Herbal cosmetics for skin and hair care. *Natural product radiance, 4*(4), 306–314.

[11]   Gupta, S. (2007). Functional foods & skin care: functional foods provide benefits beyond basic nutrition and basic skin care. *Nutraceuticals World, 10*(8), S14–S14.

[12]   Marcant, M., Lepilleur, C., Peri, E., Audibert, N., Kyer, C., Rafferty, D. W., . . . & Moran, B. (2024). Tara (Caesalpinia spinosa) Gum, a Multifunctional Polymer for Hair Care. *SOFW Journal (English version), 150*(6).

[13]   Savary, G., Grisel, M., & Picard, C. (2015). Cosmetics and personal care products. In *Natural polymers: industry techniques and applications* (pp. 219–261). Cham: Springer International Publishing.

[14]   Shiam, M. A. H., Islam, M. S., Ahmad, I., & Haque, S. S. (2025). A review of plant-derived gums and mucilages: Structural chemistry, film forming properties and application. *Journal of Plastic Film & Sheeting, 41*(2), 195–237.

[15]   Eshun, K., & He, Q. (2004). Aloe vera: a valuable ingredient for the food, pharmaceutical and cosmetic industries – a review. *Critical reviews in food science and nutrition, 44*(2), 91–96.

[16]   Fale, S. K., Umekar, M. J., Das, R., & Alaspure, M. (2022). A comprehensive study of herbal cosmetics prepared from flaxseed. *Multidiscip. Int. Res. J. Gujarat Technol. Univ, 4*, 106–112.

[17]   HE, D. B. I., & GREENS, B. T. Frontier Marshmallow Root C/S Organic, 1 lb.

[18]   Dauqan, E., & Abdullah, A. (2013). Utilization of gum arabic for industries and human health. *American Journal of Applied Sciences, 10*(10), 1270–1279.

[19]   Furtado, I. F., Sydney, E. B., Rodrigues, S. A., & Sydney, A. C. (2022). Xanthan gum: applications, challenges, and advantages of this asset of biotechnological origin. *Biotechnology Research and Innovation Journal, 6*(1), 0–0.

[20]   Mohd Haniffa, M. Z. A., Muhammad Azam, N. S. N., Syed Sulaiman, S. N., & Mohd Khamri, N. H. The effects of gums as emulsifier on the emulsion stability in cosmetic formulation.

[21]   Garti, N., & Leser, M. E. (2001). Emulsification properties of hydrocolloids. *Polymers for advanced Technologies, 12*(1-2), 123–135.

[22]   Madni, A., Khalid, A., Wahid, F., Ayub, H., Khan, R., & Kousar, R. (2021). Preparation and applications of guar gum composites in biomedical, pharmaceutical, food, and cosmetics industries. *Current Nanoscience, 17*(3), 365–379.

[23]   Malviya, R., Sharma, P. K., & Dubey, S. K. (2017). Antioxidant potential and emulsifying properties of kheri (Acacia chundra, Mimosaceae) gum polysaccharide. *Marmara Pharmaceutical Journal, 21*(3), 701–706.

[24]   Bom, S., Fitas, M., Martins, A. M., Pinto, P., Ribeiro, H. M., & Marto, J. (2020). Replacing synthetic ingredients by sustainable natural alternatives: a case study using topical O/W emulsions. *Molecules, 25*(21), 4887.

[25]   Alves, T. F., Morsink, M., Batain, F., Chaud, M. V., Almeida, T., Fernandes, D. A., . . . & Severino, P. (2020). Applications of natural, semi-synthetic, and synthetic polymers in cosmetic formulations. *Cosmetics, 7*(4), 75.

[26]   Klaschka, U. (2015). Naturally toxic: natural substances used in personal care products. *Environmental Sciences Europe, 27*(1), 1.

[27] Kumari, P. K., Akhila, S., Rao, Y. S., & Devi, B. R. (2019). Alternative to artificial preservatives. *Syst. Rev. Pharm, 10*, 99–102.

[28] Anand, S. P., & Sati, N. (2013). Artificial preservatives and their harmful effects: looking toward nature for safer alternatives. *Int. J. Pharm. Sci. Res, 4*(7), 2496–2501.

[29] Flanagan, J. (2011). Preserving cosmetics with natural preservatives and preserving natural cosmetics. *Formulating, packaging, and marketing of natural cosmetic products*, 169–178.

[30] Kumari, P. K., Akhila, S., Rao, Y. S., & Devi, B. R. (2019). Alternative to artificial preservatives. *Syst. Rev. Pharm, 10*, 99–102.

[31] Chenery, N. (2004). Organic Cosmetics for Natural Beauty!. *Organic and Natural Living*.

[32] Spaulding, A., & Baki, G. (2025). Sustainable cosmetic ingredient alternatives to replace conventional ingredients: Case studies in moisturizers and lipsticks. *International Journal of Cosmetic Science*.

[33] Liu, J. K. (2022). Natural products in cosmetics. *Natural products and bioprospecting, 12*(1), 40.

[34] Sharmeen, J. B., Mahomoodally, F. M., Zengin, G., & Maggi, F. (2021). Essential oils as natural sources of fragrance compounds for cosmetics and cosmeceuticals. *Molecules, 26*(3), 666.

[35] Burger, P., Plainfossé, H., Brochet, X., Chemat, F., & Fernandez, X. (2019). Extraction of natural fragrance ingredients: History overview and future trends. *Chemistry & biodiversity, 16*(10), e1900424.

[36] Kliszcz, A., Danel, A., Puła, J., Barabasz-Krasny, B., & Możdżeń, K. (2021). Fleeting beauty – The world of plant fragrances and their application. *Molecules, 26*(9), 2473.

[37] Frey, C. (2005). Natural flavors and fragrances: chemistry, analysis, and production.

[38] Guerra, E., Llompart, M., & Garcia-Jares, C. (2018). Analysis of dyes in cosmetics: challenges and recent developments. *Cosmetics, 5*(3), 47.

[39] PRIYA, B. S., CHITRA, K., & STALIN, T. CHAPTER ONE BIOPROSPECTING OF ECO-FRIENDLY NATURAL DYES USING NOVEL MARINE BACTERIA AS AN EFFICIENT ALTERNATIVE TO TOXIC SYNTHETIC DYES. *Microbial Biodiversity*, 1.

[40] Cui, H., Xie, W., Hua, Z., Cao, L., Xiong, Z., Tang, Y., & Yuan, Z. (2022). Recent advancements in natural plant colorants used for hair dye applications: a review. *Molecules, 27*(22), 8062.

[41] Srivastava, P., Ramesh, M., Kaushik, P., Kumari, A., & Aggarwal, S. (2022). Pyocyanin pigment from Pseudomonas species: source of a dye and antimicrobial textile finish – a review. *Proceedings of the Indian National Science Academy, 88*(4), 542–550.

[42] Ogorzałek, M., Klimaszewska, E., Małysa, A., Czerwonka, D., & Tomasiuk, R. (2024). Research on Waterless Cosmetics in the Form of Scrub Bars Based on Natural Exfoliants. *Applied Sciences, 14*(23), 11329.

[43] Talpekar, P., & Borikar, M. (2016). Formulation, development and comparative study of facial scrub using synethetic and natural exfoliant. *Research Journal of Topical and Cosmetic Sciences, 7*(1), 1–8.

[44] Pena, D. W. P., Tonoli, G. H. D., de Paula Protásio, T., de Souza, T. M., Ferreira, G. C., do Vale, I., . . . & Bufalino, L. (2021). Exfoliating agents for skincare soaps obtained from the crabwood waste bagasse, a natural abrasive from Amazonia. *Waste and Biomass Valorization, 12*(8), 4441–4461.

[45] Fernandes, C., Medronho, B., Alves, L., & Rasteiro, M. G. (2023). On hair care physicochemistry: from structure and degradation to novel biobased conditioning agents. *Polymers, 15*(3), 608.

[46] Carvalho, R. D. M., Melo, D. F., Kelati, A., & Tosti, A. (2025). With or Without Silicones? A Comprehensive Review of Their Role in Hair Care. *Skin Appendage Disorders*.

[47] Reich, C., & Su, D. T. (2001). Hair conditioners. In *Handbook of Cosmetic Science and Technology* (pp. 347–362). CRC Press.

[48] Idson, B. (2020). Polymers as conditioning agents for hair and skin. *Conditioning agents for hair and skin*, 251–279.

[49] Resende, D. I., Jesus, A., Sousa Lobo, J. M., Sousa, E., Cruz, M. T., Cidade, H., & Almeida, I. F. (2022). Up-to-date overview of the use of natural ingredients in sunscreens. *Pharmaceuticals, 15*(3), 372.

[50] BHATTACHARJEE, D., PATIL, A. B., & JAIN, V. (2021). A comparison of Natural and Synthetic Sunscreen Agents: A Review. *International Journal of Pharmaceutical Research (09752366)*, *13*(1).

[51] Dan, M. A., Ozon, E. A., HOVANEȚ, M. V., VIZITEU, H. M., Mitu, M. A., Popovici, V., . . . & Popescu, I. A. (2024). Natural sunscreens–a literature review. *carcinogenesis*, *68*, 106.

[52] Aguilera, J., Gracia-Cazana, T., & Gilaberte, Y. (2023). New developments in sunscreens. *Photochemical & Photobiological Sciences*, *22*(10), 2473–2482.

11 Challenges and Opportunities

11.1 Introduction on Regulatory and Commercialization Challenges

The role of mucilages and gums in various industries, including pharmaceuticals, food, agriculture, and cosmetics, presents both challenges and promising opportunities. One of the primary challenges lies in the regulatory and commercialization aspects. Natural polymers like mucilages and gums must meet stringent quality, safety, and efficacy standards before being approved for use, particularly in pharmaceutical and personal care products. Regulatory variations across countries further complicate global commercialization. Additionally, there are significant limitations related to scalability and cost. The extraction and purification processes for mucilages and gums can be time-consuming and expensive, especially when aiming for high purity and consistency. This limits their widespread use in large-scale industrial applications, where synthetic alternatives may still be more cost-effective [1, 3].

However, these challenges also pave the way for substantial future opportunities. Continued research and technological advancements could help optimize extraction methods, reduce production costs, and improve the functional properties of these natural polymers. Moreover, the increasing demand for sustainable, biodegradable, and plant-based alternatives is pushing industries to invest in natural sources like mucilages and gums [2]. Opportunities for research and industry collaboration are immense, especially in exploring novel plant sources, improving formulation techniques, and integrating mucilages into innovative product designs. Academic institutions, biotech firms, and regulatory agencies can work together to create standardized frameworks and support commercial viability. With a growing emphasis on green chemistry and clean-label formulations, mucilages and gums have the potential to revolutionize multiple sectors. Addressing current challenges through interdisciplinary research and industry partnerships will not only enhance their application scope but also ensure a sustainable and economically viable future for these valuable natural resources.

11.2 Regulatory Standards for Natural and Organic Cosmetics

A significant challenge in the natural and organic cosmetics industry is the lack of universally accepted regulatory standards for what qualifies as "natural" or "organic." In comparison to the food industry, where regulatory bodies like the **US Food and Drug Administration (FDA)** and **European Food Safety Authority (EFSA)** establish clear guidelines for safety and labelling, personal care products often fall into regulatory gaps, leading to inconsistencies and confusion for consumers.

https://doi.org/10.1515/9783111673509-011

11.2.1 Challenges

- **Lack of Clear Definitions**: The term "natural" remains undefined in many countries, including the United States, where the FDA does not provide a legal definition for "natural" in the context of cosmetics. This lack of clarity leaves brands to self-regulate, which can lead to misleading or vague claims about what is truly "natural" or "organic."
- **Inconsistent Certification Standards**: While certifications like **USDA Organic** and **Ecocert** offer guidance, they are not universally accepted. A product certified organic in the USA may not meet the stringent criteria for certification in the European Union or Japan. This inconsistency creates confusion for consumers who may trust one certification but find that the standards in another country differ significantly.
- **Ingredient Sourcing and Contamination Risks**: For ingredients to be considered natural or organic, they must be sourced from farms and ecosystems that adhere to strict environmental and agricultural practices. However, the risk of contamination remains a significant issue. Natural ingredients could be contaminated with synthetic chemicals, pesticides, or other harmful substances, which compromises the integrity of the product and the certification process.

11.2.2 Potential Solutions

- **Developing Unified Global Standards**: One potential solution is the establishment of clearer, globally recognized guidelines for what constitutes "natural" or "organic" cosmetics. Organizations like **COSMOS-standard** and **EcoCert** are already working to provide more unified criteria, but their reach must be expanded to ensure global acceptance.
- **Third-Party Certifications**: Companies should strive for third-party certifications such as **COSMOS**, **Ecocert**, or the **Leaping Bunny** certification for cruelty-free products. These certifications establish a level of trust between brands and consumers, helping to ensure that products are genuinely natural or organic, and free of harmful ingredients.
- **Clearer Labeling**: Brands should be encouraged to label products with more transparency, clearly stating whether an ingredient is organic, wildcrafted, or cultivated using sustainable farming practices. These labels would help consumers make more informed choices and foster greater trust in the industry.

11.3 Safety and Toxicity Concerns in Natural Ingredients

While natural ingredients are often marketed as safer alternatives to synthetic chemicals, many plant-based ingredients still carry safety risks, especially if not used correctly. Some natural ingredients such as essential oils, plant extracts, and herbal formulations may cause allergic reactions, irritation, or toxicity if not carefully sourced, formulated, or processed.

11.3.1 Challenges

- **Unforeseen Reactions**: Even though an ingredient may be deemed safe in one formulation, its effects can change when combined with other ingredients or used at high concentrations. For example, **tea tree oil** or **citrus oils** are known to cause skin irritation or photosensitivity, which can lead to sunburn or skin discoloration in some individuals.
- **Lack of Long-Term Testing**: Many natural ingredients have not undergone comprehensive long-term safety testing in the context of personal care products. While some plants have been used for centuries in traditional medicine, there may still be risks associated with prolonged or excessive exposure to these ingredients.
- **Ingredient Interactions**: The complexity of natural formulations means that ingredients can interact in unpredictable ways. Some oils, for example, can destabilize emulsions (blended mixtures of water and oil), leading to product failure or even skin irritation when applied.

11.3.2 Potential Solutions

- **Comprehensive Testing**: Manufacturers should conduct extensive clinical trials, dermatological testing, and **patch tests** to ensure the safety of their formulations. Additionally, testing should be done across a range of skin types and sensitivities to ensure that natural products can be used safely by a wide variety of consumers.
- **Formulation Transparency**: Transparent labeling is essential. Companies must include detailed information about the source, concentration, and origin of ingredients, helping consumers understand the potential risks and benefits of each component.
- **Collaborating with Experts**: Brands should work with dermatologists, toxicologists, and chemists to ensure that formulations are safe and effective. This will ensure that products do not cause adverse reactions, especially when used in conjunction with other natural or synthetic products.

11.4 Compliance with International Trade Regulations

Companies aiming to expand into global markets must ensure that their products comply with the regulatory requirements of each target country. This is a significant challenge, as cosmetic regulations and ingredient restrictions vary widely across borders, creating complexity for companies navigating international markets.

11.4.1 Challenges

- **Inconsistent Labeling Regulations**: Different countries have different rules regarding labeling claims for personal care products. For instance, the **European Union (EU)** has strict regulations on labeling claims like "organic" or "hypoallergenic," while the **United States** has more relaxed guidelines, leading to potential confusion for consumers and brands.
- **Cross-Border Ingredient Approval**: Ingredients that are approved in one country may be banned or restricted in another. For example, **citrus extracts** are restricted in some countries due to concerns about photosensitivity, yet in others, they are used widely in cosmetics.
- **Testing and Animal Welfare Standards**: Even when ingredients are natural, the regulatory environment concerning **animal testing** is strict in many regions, such as the EU, which mandates that all cosmetic products must be tested without animal cruelty. Companies must ensure that their ingredients and final products meet these requirements, which can complicate market entry.

11.4.2 Potential Solutions

- **Global Regulatory Experts**: Companies can benefit from collaborating with experts who specialize in global cosmetic regulations. These experts can help ensure compliance with local and international laws, allowing companies to tailor their formulations, packaging, and marketing strategies according to each country's unique requirements.
- **Centralized Compliance Hub**: Establishing a centralized team to monitor and respond to the evolving regulatory landscape across multiple markets would help companies ensure consistent compliance. This team can keep track of changing regulations and ensure that product formulations, ingredient sourcing, and labeling practices align with local requirements.
- **Adapting to Local Regulations**: When entering foreign markets, brands must adapt their formulations and marketing messages to comply with the regulations of that country. This may involve adjusting ingredient lists, modifying product claims, or even changing packaging to comply with local standards.

11.5 Consumer Misconceptions and Education

One of the biggest challenges for natural and organic cosmetics is addressing consumer misconceptions about the meaning and safety of "natural" products. While many consumers associate "natural" ingredients with safety and effectiveness, they may be unaware that not all natural ingredients are suitable for every skin type, or that certain ingredients may still cause side effects.

11.5.1 Challenges

- **Lack of Consumer Education**: Consumers are often confused by the broad range of terms used in natural cosmetics, such as "organic," "green," "eco-friendly," or "pure." This lack of education makes it difficult for them to distinguish between truly natural products and those that are labeled misleadingly.
- **Greenwashing**: Greenwashing is a deceptive marketing tactic where companies make unsubstantiated claims about their products being natural or eco-friendly without providing proof or certifications. This not only harms consumer trust but also undermines the integrity of the entire natural personal care industry.
- **Cost vs. Value**: Many consumers perceive natural products as more expensive, but they may not fully understand the added value these products offer in terms of safety, efficacy, and environmental sustainability. This can hinder the growth of the market.

11.5.2 Potential Solutions

- **Transparency in Marketing**: Brands should embrace transparency in their marketing strategies, offering clear information about ingredient sourcing, benefits, and potential risks. This can help educate consumers and foster greater trust in the brand.
- **Third-Party Certifications**: Independent certifications such as **USDA Organic**, **Ecocert**, and **COSMOS** offer consumers a reliable indication of the product's natural or organic status. Displaying these certifications on packaging can help to reduce confusion and validate claims.
- **Incorporating Education into Branding**: Personal care brands should include educational content as part of their branding efforts. This could include social media campaigns, website content, or in-store displays that inform consumers about the benefits of natural ingredients, the importance of ethical sourcing, and the potential risks of unverified claims.

11.6 Sourcing and Sustainability Challenges

Ethical sourcing and sustainability are central to the natural personal care industry. As demand for natural ingredients grows, the challenge in ensuring that these raw materials are harvested responsibly becomes more pressing.

11.6.1 Challenges

- **Sustainability and Overharvesting**: Some natural ingredients like **sandalwood** and **wildcrafted herbs** are at risk of overharvesting, which can lead to environmental degradation and loss of biodiversity.
- **Fair Trade and Labor Practices**: Ensuring that the labor involved in sourcing and harvesting ingredients is ethical is vital. Unfortunately, the supply chain for many natural ingredients is opaque, making it difficult for companies to ensure that workers are treated fairly and that communities benefit from trade.
- **Carbon Footprint of Sourcing**: Although natural ingredients are often considered eco-friendly, the environmental impact of sourcing and transporting these ingredients can be significant. For example, raw materials that are sourced from remote areas can contribute to a high carbon footprint, especially if transported by air or long distances.

11.6.2 Potential Solutions

Sustainable Sourcing Partnerships: Brands can build partnerships with local farmers and cooperatives that focus on sustainable harvesting and fair trade practices. These partnerships not only ensure that ingredients are sourced in environmentally responsible ways but also help to improve the livelihoods of communities involved in harvesting.

Sustainability Certifications: Working with recognized sustainability certification organizations such as **Fair Trade** or **Rainforest Alliance** can help ensure that ingredients are harvested ethically and sustainably.

Circular Economy Models: Brands can adopt circular economy principles, focusing on reducing waste, recycling materials, and minimizing the carbon footprint associated with sourcing. This can include packaging that is biodegradable or refillable, as well as reducing the use of harmful chemicals during manufacturing.

11.7 Overcoming Limitations in Scalability and Cost [3]

11.7.1 High Production Costs of Natural Ingredients

Natural ingredients often come with higher upfront production costs compared to their synthetic counterparts. The extraction, processing, and refinement of plant-based compounds can be labor-intensive and require specialized equipment. Additionally, the cost of raw materials may fluctuate based on seasonal availability, supply chain constraints, and geopolitical factors. These factors can make scaling production for natural ingredients more expensive.

11.7.2 Challenges

– **Raw Material Sourcing:** Many natural ingredients are sourced from specific geographical regions, which may limit supply and increase costs due to transportation, tariffs, and limited access to high-quality raw materials.
– **Extraction and Processing:** Extracting beneficial compounds from plants, minerals, or other natural sources is often a complex and energy-intensive process. The quality of the end-product depends heavily on the methods of extraction, and scaling these methods can be expensive.
– **Sustainability Concerns:** While natural ingredients are considered environmentally friendly, overharvesting and unsustainable practices can increase costs. Companies must often balance sustainable sourcing with cost-effectiveness, a challenge when demand outpaces supply.

11.7.3 Potential Solutions

– **Vertical Integration:** Companies can gain greater control over costs and scalability by vertically integrating their supply chains. This involves directly controlling the farming, extraction, and manufacturing processes rather than relying on third-party suppliers. By owning or partnering with farms, businesses can secure a steady, sustainable supply of natural ingredients at lower costs and reduce the risk of price fluctuations.
– **Investing in Efficient Extraction Technologies:** Innovations in **extraction technology** such as **cold-press extraction, supercritical CO_2 extraction**, and **enzyme-assisted extraction** can help streamline the process, reduce energy consumption, and lower production costs. By adopting more efficient technologies, manufacturers can produce natural ingredients at a larger scale without sacrificing quality.

– **Incentivizing Sustainable Practices:** Companies can collaborate with sustainable agriculture initiatives and create partnerships with farmers to ensure that raw materials are sourced in a way that is both cost-effective and ecologically responsible. Long-term contracts with farmers or cooperatives can stabilize prices and make raw material sourcing more predictable.

11.8 Supply Chain Constraints

Scalability of natural ingredients is often constrained by **supply chain limitations**. The agricultural and botanical nature of many natural personal care ingredients means that production depends on factors such as crop yields, seasonal variations, and climate conditions. Disruptions in the supply chain – whether due to poor harvests, extreme weather, or geopolitical tensions – can lead to shortages, delays, and price increases, making it challenging to scale production to meet growing demand.

11.8.1 Challenges

– **Geographical Constraints:** Many natural ingredients such as **argan oil**, **shea butter**, and **coconut oil** are only grown in specific regions. This can create bottlenecks in the supply chain and result in limited availability of raw materials.
– **Climate Vulnerability:** Extreme weather events such as droughts, floods, or storms can damage crops and reduce yields. Climate change is increasingly threatening the stability of supply chains, particularly for ingredients that require stable growing conditions.
– **Dependence on Small-Scale Producers:** In many cases, natural ingredient production is carried out by small-scale farmers or cooperatives. This can limit the scalability of operations, as small producers may not have the capacity to meet large-scale commercial demands or adhere to standardized production processes.

11.8.2 Potential Solutions

– **Diversified Sourcing:** To mitigate the risk of supply chain disruptions, companies can diversify their sourcing strategies. By cultivating relationships with multiple suppliers in different regions or countries, they can reduce dependence on any single geographic area or producer, ensuring a more consistent supply of raw materials.
– **Agroforestry and Crop Rotation:** Encouraging the use of agroforestry systems and crop rotation can increase the resilience of agricultural production and protect against climate risks. In these systems, multiple crops are grown together to

maximize yield, improve soil health, and reduce the impact of pests or extreme weather.
- **Supporting Small-Scale Farmers:** Large-scale manufacturers can establish direct relationships with small-scale farmers to help them scale up production. Through training, technology transfer, and fair trade agreements, farmers can be empowered to increase yields, improve product quality, and meet commercial demands more efficiently.

11.9 High Consumer Price Sensitivity

Despite the clear benefits of natural personal care products, **price sensitivity** remains a significant barrier to widespread adoption. Natural ingredients often come with a premium price tag due to the costs associated with sourcing, extraction, and production. Consumers who are accustomed to more affordable synthetic alternatives may be reluctant to switch to pricier natural products, especially when the cost difference is substantial.

11.9.1 Challenges

- **Premium Pricing:** Because natural ingredients are often more expensive to produce, products containing them typically carry a higher retail price. This price difference may deter price-conscious consumers who view natural products as a luxury rather than a necessity.
- **Value Perception:** Even when natural products are priced higher, some consumers may not perceive the added value in terms of benefits, especially if they are unfamiliar with the efficacy of natural ingredients.
- **Competing with Synthetic Alternatives:** Many synthetic chemicals used in personal care products are mass-produced and relatively inexpensive, creating strong competition for natural personal care items.

11.9.2 Potential Solutions

- **Educating Consumers on Value:** Effective education campaigns can help consumers understand the **long-term value** of natural products such as their **health benefits, sustainability**, and **lack of harmful chemicals**. Communicating these benefits through clear, transparent marketing can justify the price premium.
- **Bulk Production and Economies of Scale:** By increasing production volumes, companies can achieve **economies of scale**, which helps reduce the per unit cost

of natural products. In the long term, this can help lower prices and make natural personal care products more accessible to a broader audience.
- **Product Innovation and Packaging Optimization:** Manufacturers can innovate in packaging and formulation to reduce costs. For example, creating concentrated formulas that require smaller packaging sizes can help lower shipping and packaging costs. Similarly, offering refillable packaging options may reduce both the price point for consumers and the environmental impact of the product.

11.10 Formulation and Ingredient Compatibility

Scaling natural ingredients presents challenges in **formulation** – combining natural raw materials into stable, safe, and effective products. Many natural ingredients are more sensitive to temperature, light, and air exposure, and their properties can degrade more quickly than synthetic ingredients. Furthermore, some natural compounds may not mix well with others, leading to issues with texture, viscosity, or shelf life.

11.10.1 Challenges

- **Stability Issues:** Natural ingredients such as **essential oils** or **plant extracts** may be more prone to oxidation, degradation, or microbial growth, which can impact the stability and shelf life of the product. This can make it difficult to scale production without compromising product quality.
- **Formulation Limitations:** Natural ingredients may not always provide the same functional properties as synthetic chemicals. For example, natural preservatives might not be as effective in preventing bacterial growth as synthetic ones, necessitating additional research and development to find suitable alternatives.
- **Compatibility Issues:** Some natural ingredients may not mix well with others, leading to problems like separation, clumping, or instability in emulsions.

11.10.2 Potential Solutions

- **Research and Innovation in Natural Stabilizers:** As consumer demand for natural products grows, investment in research on natural stabilizers, preservatives, and emulsifiers can help address formulation challenges. Natural alternatives such as **rosemary extract**, **grapefruit seed extract**, and **vitamin E** can extend shelf life without the need for synthetic chemicals.
- **Advanced Formulation Technologies:** Companies can invest in technologies such as **nano-encapsulation** or **microencapsulation** to improve the delivery

and stability of natural ingredients in personal care products. These technologies can protect sensitive ingredients from oxidation, improve their shelf life, and enhance their efficacy.

–   **Formulation Trials:** Extensive testing and trials are necessary to ensure that natural ingredients maintain their integrity and performance across various formulations. Companies can work closely with formulation scientists to create stable, effective products that meet consumer expectations.

11.11 Scaling Ethical and Sustainable Practices

As the demand for natural personal care ingredients increases, maintaining **ethical and sustainable practices** becomes increasingly difficult. Ensuring fair trade, environmentally friendly sourcing, and sustainable production at scale requires careful planning, collaboration, and transparency.

11.11.1 Challenges

–   **Sustainable Sourcing at Scale:** The larger the scale, the greater the pressure on ecosystems and communities. Scaling up natural ingredient production in an environmentally responsible way requires sustainable farming practices, fair labor conditions, and a commitment to biodiversity.
–   **Transparency in Sourcing:** As the market expands, maintaining transparency across the supply chain becomes increasingly challenging. Companies must ensure that they are sourcing ingredients ethically and that all stakeholders (farmers, workers, and manufacturers) are treated fairly.

11.11.2 Potential Solutions

Collaborative Partnerships for Sustainability: Brands can form partnerships with sustainable farming initiatives, certified ethical organizations, and nongovernmental organizations (NGOs) that focus on improving the environmental and social impact of agriculture. These partnerships can help ensure that scaling up production does not compromise sustainability.

Tracking and Traceability: Implementing advanced traceability systems that track the journey of ingredients from farm to factory can help companies ensure compliance with sustainable and ethical practices. Blockchain technology, for example, is increasingly being used to track the sourcing of ingredients, ensuring transparency and accountability.

11.12 Future Opportunities for Research and Industry Collaboration [4]

The search for novel sources of mucilages and gums is an exciting area of research in the natural cosmetics sector. Many indigenous plants and lesser-known species may hold the key to more sustainable, efficient, and innovative ingredients. These ingredients could offer specialized benefits that are not yet fully realized in the market.

11.12.1 Research Opportunities

- **Identifying Novel Plant Sources**: Historically, mucilages and gums have been derived from well-known plants like **guar gum**, **xanthan gum**, and **acacia gum**. However, there remains a wealth of untapped plant species in biodiverse ecosystems, particularly in tropical forests, arid deserts, and high-altitude regions. By investigating indigenous plants in these ecosystems, researchers could discover new species with unique mucilaginous properties. For example, **moringa** and **baobab** are two such plants that have been traditionally used for their beneficial properties but may also offer undiscovered mucilage compounds with specialized benefits.
- **Tropical Forests**: These environments are rich in plant species with highly adaptive properties. Exploring plants like **coconut** and **aloe Vera** has already yielded valuable mucilaginous substances, but further research could lead to more species with enhanced moisture-retention properties or even specific benefits for sensitive skin.
- **Arid Deserts**: Some desert plants, such as **cactus** and **gum arabic**, are known for their ability to survive in harsh conditions. These plants' mucilages could possess high adaptability, which could be useful for addressing skin concerns like dehydration or environmental stress.
- **Sustainability and Sourcing**: As the demand for natural ingredients increases, it is essential to ensure that mucilage and gum production does not lead to overharvesting and environmental depletion. Research into sustainable farming practices, such as **agroforestry** and the domestication of specific mucilaginous plants, could help create a reliable supply chain while preserving ecosystems.
- **Agroforestry Systems**: Growing mucilaginous plants alongside other crops in agroforestry systems can promote biodiversity and reduce environmental degradation. This method ensures that the ecosystem remains intact, while the plants are cultivated in a way that supports both environmental and social sustainability.
- **Cultivation of Specific Crops**: Developing farming methods for mucilaginous plants like **aloe vera**, **moringa**, or **okra** could enable more controlled, efficient production without depleting natural resources. Creating cultivation methods for

previously wild-harvested plants reduces pressure on natural ecosystems and promotes economic growth in rural areas.

11.13 Industry Collaboration

- **Partnerships with Ethnobotanists and Indigenous Knowledge Holders**: Indigenous communities have long used plants for medicinal and cosmetic purposes, but their knowledge is often overlooked. Collaborating with ethnobotanists and local communities can uncover novel sources of mucilage and gums that have been used for centuries but are not yet commercially exploited. These partnerships are critical to discovering plant species with beneficial properties and ensuring that indigenous communities are compensated for their knowledge.
- **Collaboration with Agricultural Innovators**: Working with agricultural research centers and companies focused on the domestication of wild plant species can create sustainable, efficient supply chains for mucilage and gum ingredients. This partnership would also help optimize farming techniques, improve crop yields, and increase the efficiency of production processes for the beauty industry.

11.13.1 Functionalizing Mucilages and Gums for Targeted Benefits

While mucilages and gums are known for their basic moisturizing and emulsifying properties, there is a growing interest in functionalizing these natural ingredients to provide more specific and targeted benefits for skin and hair. The beauty industry is increasingly looking for active ingredients that can deliver more than just hydration but also antiaging, anti-inflammatory, and skin-healing benefits.

11.13.2 Research Opportunities

- **Bioactive Compounds**: Mucilages and gums are rich in polysaccharides and other bioactive compounds that can offer additional therapeutic effects. **Isolating bioactive compounds** like antioxidants, anti-inflammatory agents, and growth factors from mucilages could enhance their efficacy in addressing specific skin concerns. For example:
- **Antiaging**: Combining mucilages with compounds that promote collagen synthesis or reduce oxidative stress could lead to the development of advanced antiaging products. These compounds could help to reduce wrinkles, increase skin elasticity, and promote skin regeneration.

- **Anti-Inflammatory**: Some mucilages and gums, such as those derived from **marshmallow root** or **slippery elm**, may contain anti-inflammatory properties that can soothe irritated skin or reduce redness, making them ideal for sensitive skin or conditions like acne and rosacea.
- **Synergy with Other Natural Ingredients**: Research into the synergy between mucilages and other natural ingredients (like essential oils, herbal extracts, and vitamins) could lead to highly effective formulations. Combining mucilages with ingredients like **vitamin C** (for brightening) or **hyaluronic acid** (for deep hydration) could create powerful skin care products targeting specific concerns like dry skin, fine lines, or uneven skin tone.

11.13.3 Industry Collaboration

- **Collaboration with Research Institutions and Universities**: Partnering with academic institutions to conduct in-depth studies on the bioactive compounds found in mucilages and gums could accelerate the development of specialized formulations. University collaborations can provide valuable insights into how mucilages can interact with other active ingredients, leading to optimized formulations for specific skin or hair concerns.
- **Collaborating with Biotechnology Companies**: Biotechnology companies can help improve the **bioavailability** of compounds found in mucilages and gums. By using biotechnology to enhance the absorption of active compounds or creating more stable and potent versions, these companies could help boost the therapeutic effects of mucilage-based formulations, leading to more innovative and effective products in the market.

11.14 Personalization and Customization of Skin Care and Hair Care Products

The demand for personalized beauty products has increased significantly, driven by consumer preferences for products tailored to their specific skin type, concerns, or environmental factors. Mucilages and gums, with their ability to adapt to varying moisture levels and soothe the skin, are ideal ingredients for personalized formulations.

11.14.1 Research Opportunities

- **Smart Formulation Development**: One exciting research area is the development of "smart" mucilage-based formulas that respond to changes in the skin's or hair's moisture levels. These formulations could adjust their properties based on environmental conditions (e.g., humidity or temperature) or the skin's specific needs, offering consumers a more tailored experience. **Smart moisturizers** could be developed to release active ingredients based on real-time skin hydration levels, ensuring more precise and personalized care.

- **Microbiome-Friendly Products**: Another promising research area is understanding how mucilages and gums interact with the skin or scalp microbiome. Products that nourish and balance the **microbiome** (the collection of beneficial bacteria on the skin) could lead to healthier skin or hair by promoting the growth of beneficial bacteria while suppressing harmful ones. Research into how mucilage can support microbiome health could result in a new generation of **microbiome-friendly** products, benefiting consumers with sensitive skin, acne, or dandruff.

11.14.2 Industry Collaboration

- **Consumer Data Integration**: Collaborating with companies that specialize in **personalized skin care** platforms (e.g., AI-driven technology) could allow for the creation of custom formulas based on individual consumer needs. These platforms can assess factors such as moisture levels, skin concerns, and environmental conditions to create a personalized skin care regimen that includes mucilage-based products.
- **Tailored Products for Diverse Demographics**: Understanding how different demographics react to mucilage- and gum-based formulations is essential for the development of region-specific products. For example, individuals living in humid, tropical climates may require more lightweight formulations that hydrate without causing greasy buildup, while those in dry, desert climates might need richer, more emollient formulas.

11.15 Improving Stability and Delivery Mechanisms

One of the key challenges with mucilages and gums is maintaining their stability in cosmetic formulations and optimizing their delivery to the skin or hair. These natural ingredients are often prone to degradation over time, and they may not always penetrate the skin effectively.

11.15.1 Research Opportunities

- **Formulation Stability**: Mucilages and gums can be sensitive to temperature, pH, and environmental conditions, which may lead to degradation or changes in their properties. Research into how mucilages interact with other ingredients and how their properties evolve over time could improve their stability. Advanced **preservation techniques** or natural stabilizers could help extend the shelf life of mucilage-based products, making them more commercially viable.

- **Encapsulation Technologies**: **Encapsulation** techniques such as liposomes or nanoparticles could be used to protect mucilage molecules and increase their penetration into the skin or hair. These delivery systems can help improve the **bioavailability** of active compounds and ensure that they are released gradually over time for sustained benefits. This technology could be especially useful for **antiaging** or **anti-inflammatory** products that require long-term effects.

11.15.2 Industry Collaboration

- **Partnerships with Delivery System Innovators**: Working with companies specializing in **advanced delivery systems** could help overcome the challenge of ingredient stability and absorption. Innovations in liposomal or nanoparticle technology could enhance the effectiveness of mucilage-based products by ensuring that they penetrate the skin more deeply and release their active compounds over a longer period.
- **Collaboration with Cosmetic Chemists**: Experienced formulation scientists can provide valuable insight into the methods to combine mucilage with other natural preservatives and stabilizers to create more effective, shelf-stable products. Such collaborations could result in mucilage-based formulations that retain their efficacy over time without relying on synthetic chemicals.

11.16 Sustainable and Ethical Practices in Mucilage and Gum Sourcing

As the beauty industry continues to grow, so does consumer demand for ethically sourced and sustainable ingredients. Mucilages and gums, being natural products, are often marketed as eco-friendly, but sourcing practices must be carefully managed to ensure that they do not contribute to environmental degradation or unfair labor practices.

11.16.1 Research Opportunities

- **Carbon Footprint of Mucilage and Gum Production**: Understanding the environmental impact of mucilage and gum production is essential for making these ingredients more sustainable. Researching ways to reduce carbon emissions associated with harvesting and processing mucilage-based ingredients will help brands meet the demands of eco-conscious consumers.

- **Circular Economy Approaches**: Investigating ways to reuse by-products from mucilage and gum extraction could contribute to a circular economy model in the beauty industry. Reducing waste and ensuring that every part of the plant is used in some way could significantly increase sustainability in the production process.

11.16.2 Industry Collaboration

- **Partnerships with Sustainability Organizations**: Collaborating with organizations focused on **sustainable agriculture** and **ethical sourcing** can help ensure that mucilage and gum ingredients are harvested in a way that supports environmental conservation and promotes fair labor practices.
- **Eco-friendly Packaging Innovations**: Collaborating with packaging innovators to develop **biodegradable** or **recyclable** packaging for mucilage and gum-based products could further strengthen a brand's commitment to sustainability and eco-friendliness.

References

[1]   de Carvalho Coelho, L. M., Veloso, B. F., de Lima Silva, V., Assunção, L. S., Ribeiro, C. D. F., & Otero, D. M. (2025). Innovations and challenges in the use of cactus mucilage: A technological analysis. *International Journal of Biological Macromolecules*, 144792.

[2]   Tosif, M. M., Najda, A., Bains, A., Kaushik, R., Dhull, S. B., Chawla, P., & Walasek-Janusz, M. (2021). A comprehensive review on plant-derived mucilage: characterization, functional properties, applications, and its utilization for nanocarrier fabrication. *Polymers*, *13*(7), 1066.

[3]   Shishir, M. R. I., Suo, H., Taip, F. S., Ahmed, M., Xiao, J., Wang, M., . . . & Cheng, K. W. (2024). Seed mucilage-based advanced carrier systems for food and nutraceuticals: fabrication, formulation efficiency, recent advancement, challenges, and perspectives. *Critical Reviews in Food Science and Nutrition*, *64*(21), 7609–7631.

[4]   Corrales-García, J. (2007, October). Industrialization of cactus pads and fruit in Mexico: Challenges and perspectives. In *VI International Congress on Cactus Pear and Cochineal 811* (pp. 103–112).

12 Conclusion: The Future of Mucilage and Gums in Sustainable Applications

12.1 Summary of Advancements and Findings

In the world of beauty and personal care, mucilages and gums have long been appreciated for their hydrating, emulsifying, and thickening properties. These natural ingredients are derived from plants and have gained popularity due to their eco-friendly credentials. However, advancements in research and sustainability practices have expanded their potential applications.

12.1.1 Advancement: Identification of New, Under-Researched Plant Sources

As traditional sources like guar gum, xanthan gum, and acacia gum become more commercially available, the search for new plant species with mucilaginous properties has accelerated. Tropical ecosystems and indigenous plant species, in particular, hold significant potential for novel mucilage sources. Plants like baobab, *moringa*, and okra, which were previously underutilized, are now being explored for their potential to offer enhanced or unique mucilaginous properties. This is especially important as the demand for natural ingredients continues to grow.

- **Indigenous Plants:** Many indigenous plants have long been used by local communities for their medicinal and cosmetic benefits. For example, the Indian gooseberry (*amla*) has been used in Ayurvedic medicine for its antioxidant and hydrating properties, and its mucilage could be useful for skin hydration or anti-aging formulations.
- **Tropical Plants:** Plants from tropical forests, like the mango or papaya, produce mucilage that can be harvested sustainably and used as a natural emulsifier. Their mucilage is often rich in beneficial bioactive compounds, making them not only useful for hydrating the skin but also potentially offering antioxidant and anti-inflammatory benefits.

12.1.2 Finding: Sustainable Mucilage Sources Through Regenerative Agricultural Practices

Regenerative agriculture practices are gaining attention for their ability to enhance soil health, increase biodiversity, and ensure that plants are grown in a way that minimizes environmental damage. This approach can be extended to mucilage-producing plants to create a more sustainable supply chain. Regenerative farming methods like no-till farming, cover cropping, and crop rotation can be applied to the cultivation of

https://doi.org/10.1515/9783111673509-012

mucilage-producing plants, reducing the environmental impact and ensuring that the plants are grown in a manner that replenishes the soil and ecosystem.

By incorporating regenerative practices, mucilage and gum-producing plants like okra, aloe vera, and acacia can be grown in diverse ecosystems without depleting natural resources. This not only provides a renewable and sustainable supply of these key ingredients but also contributes positively to local farming communities by promoting sustainable livelihoods.

12.2 Sustainability in Harvesting and Production of Mucilages and Gums

With the beauty and personal care industry becoming more environmentally conscious, the push towards sustainable harvesting and production methods for mucilage and gum-producing plants is paramount. This includes utilizing agroforestry systems and sustainable monocropping methods to prevent the depletion of natural resources.

12.2.1 Advancement: Focus on Sustainable Farming Techniques

Agroforestry, which involves growing multiple plant species in a mutually beneficial manner, offers a promising way to cultivate mucilaginous plants sustainably. Instead of monocropping, which can lead to soil degradation and biodiversity loss, agroforestry systems encourage the growth of a variety of crops, leading to healthier ecosystems and improved soil fertility.

For example, growing acacia trees alongside other crops such as coriander or cumin can create a balanced ecosystem that protects the soil and reduces the need for synthetic fertilizers. This reduces the carbon footprint of mucilage production while promoting biodiversity.

Similarly, sustainable monocropping techniques are evolving, where mucilage-producing plants like guar are cultivated in such a way that soil health is maintained, and irrigation requirements are minimized. By using integrated pest management (IPM) and organic farming methods, mucilage and gum plants can be grown in a way that does not harm the environment or deplete resources.

12.2.2 Finding: Plant-Based Gums Can Be Sourced Sustainably

By adopting sustainable farming techniques, it is possible to source plant-based gums without causing ecological harm. For example, guar gum, often used in shampoos and lotions, can be grown with minimal impact on the ecosystem, provided it is cultivated through sustainable methods like rain-fed irrigation and crop rotation. These practices ensure that the land can be used repeatedly without degrading soil health.

In addition, Xanthan gum, another popular natural gum used in personal care formulations, can be derived through fermentation processes. These processes can be optimized to require fewer resources, such as water and energy, compared to traditional cultivation methods, thus offering an eco-friendly alternative to synthetic ingredients.

12.3 Biodegradable and Eco-friendly Alternatives

As the beauty industry increasingly embraces sustainability, the use of biodegradable and eco-friendly ingredients has become a top priority. Mucilages and gums, derived from plants, provide a natural alternative to petrochemical-based ingredients that can be harmful to the environment.

12.3.1 Advancement: Biodegradable Emulsifiers from Mucilages and Gums

Mucilages and gums, such as guar gum, xanthan gum, and gellan gum, have excellent emulsifying properties. These ingredients help stabilize products by allowing oil and water to mix, and they can be used as biodegradable emulsifiers in formulations such as lotions, shampoos, and creams. By replacing petrochemical-based emulsifiers like propylene glycol or sodium lauryl sulfate, mucilage-based formulations contribute to reducing pollution and environmental toxicity.

These natural emulsifiers are not only biodegradable but also have a much lower environmental impact compared to their synthetic counterparts. For instance, guar gum is produced through sustainable farming practices, and its extraction and processing require minimal energy compared to the production of synthetic emulsifiers.

12.3.2 Finding: Eco-friendly Stabilizers and Reduced Environmental Toxicity

The use of natural gums in personal care products significantly reduces the overall environmental toxicity of the products. Xanthan gum, for instance, is widely used in personal care formulations as a stabilizer, and its biodegradable nature ensures that it does not contribute to long-term pollution in waterways. Additionally, gellan gum, often used as a thickening agent, is derived from fermentation processes, making it an eco-friendly alternative that is produced with a minimal carbon footprint.

The demand for eco-friendly stabilizers is growing, and mucilage-based gums are poised to play a crucial role in reducing the reliance on harmful petrochemical-based ingredients in the beauty industry. Brands that embrace these alternatives can position themselves as leaders in sustainable beauty, responding to the needs of environmentally conscious consumers.

12.4 Low Carbon Footprint Manufacturing

The beauty industry is increasingly focused on reducing its carbon footprint, and this extends to the production of natural ingredients such as mucilages and gums. Advancements in extraction methods and sustainable farming practices have made it possible to produce mucilage and gum-based ingredients with a significantly lower carbon footprint.

12.4.1 Advancement: Energy-Efficient Extraction Methods

Research into more energy-efficient extraction methods for mucilages and gums has led to improvements in the efficiency of production processes. For example, traditional extraction methods for guar gum involved boiling or drying the plant material, which required significant amounts of energy. However, modern techniques, such as cold extraction and supercritical fluid extraction, have been developed, which require less energy and reduce the environmental impact.

Additionally, water usage in mucilage extraction is being optimized. By utilizing closed-loop water systems and water recycling technologies, the amount of water required for extraction can be reduced, helping to preserve local water resources and minimize the environmental footprint of mucilage production.

12.4.2 Finding: More Sustainable Manufacturing Practices

Mucilage and gum extraction processes, such as those used to derive okra or aloe vera gum, have been optimized to minimize water and energy usage. These changes, alongside improved transportation methods that reduce carbon emissions, contribute to making mucilage-based personal care products more sustainable in terms of both production and distribution.

For example, aloe vera plants, known for their mucilage-rich properties, can now be grown in water-scarce regions using efficient irrigation techniques, further reducing the carbon footprint of aloe vera-based personal care products. Similarly, okra, which is cultivated primarily in warm climates, can be harvested with minimal environmental impact if it is grown using sustainable agricultural methods.

12.5 Waste Reduction and Circular Economy

The circular economy model, which focuses on reducing waste and reusing materials, is gaining traction in industries around the world. The beauty industry, too, is looking for ways to integrate circular economy principles into the production of personal care products, including those based on mucilages and gums.

12.5.1 Advancement: Use of Waste Products from Mucilage Extraction

In the past, plant residues from mucilage extraction, such as the leaves, stems, and seeds, were often discarded as waste. However, with advancements in circular economy practices, many of these by-products are now being repurposed for value-added applications. For example, the leftover okra plant material can be used as animal feed or converted into biodegradable packaging materials.

The use of these by-products not only reduces waste but also creates new revenue streams for farmers and businesses. In this way, mucilage production contributes to a more sustainable and circular economy by ensuring that every part of the plant is utilized.

12.5.2 Finding: Waste Products Contribute to a Circular Economy

The concept of zero-waste production is becoming increasingly common in the cosmetics industry. By utilizing waste products from mucilage and gum extraction, companies can significantly reduce their environmental impact. For instance, xanthan gum production often leaves behind plant residues, which can be repurposed into biodegradable plastics or organic fertilizers.

This innovative approach not only minimizes the environmental footprint of mucilage and gum production but also promotes a more sustainable business model for beauty brands.

12.6 The Role of Mucilage and Gums in Achieving Sustainability Goals

12.6.1 Reduction of Environmental Impact Through Biodegradability

One of the core goals of sustainability is reducing environmental pollution, especially from synthetic chemicals and nonbiodegradable substances that accumulate over time. Mucilages and gums are a sustainable alternative because they are **biodegradable** and break down quickly in the environment, unlike many synthetic polymers.

12.6.2 Goal: Reducing Environmental Pollution from Synthetic Chemicals

Synthetic chemicals and polymers are a major contributor to environmental pollution, particularly in the personal care and cosmetic industries. Many of these synthetic substances take hundreds of years to degrade in landfills and wastewater systems. Mucilages and gums, however, offer a significant advantage as biodegradable alternatives, helping mitigate the problem of long-term environmental contamination.

12.6.3 Data

-   **Xanthan Gum:** This commonly used gum in cosmetics and food products has a biodegradation rate of around 90% within 2–4 weeks in wastewater. This is in stark contrast to synthetic polymers, which can take up to 100 years to degrade, causing significant environmental damage.
-   **Guar Gum:** Known for its hydrating properties, guar gum is also biodegradable, with a degradation rate of approximately 80–90% over a few weeks. The high biodegradation rates mean that products containing guar gum break down quickly, reducing their environmental impact.
-   **Gellan Gum:** Another popular gum used in cosmetics and food, gellan gum is 85–95% biodegradable in 2–4 weeks. This rate significantly lowers the environmental toxicity and supports waste reduction initiatives.

12.6.4 Impact

The use of mucilages and gums in place of nonbiodegradable ingredients helps reduce overall environmental pollution (Table 12.1). By replacing synthetic polymers with biodegradable options, industries can significantly reduce the long-term environmental damage caused by their products.

Table 12.1: Environmental impact of mucilage.

Ingredient	Biodegradation rate	Environmental impact
Xanthan gum	90% in 2–4 weeks	Low environmental toxicity
Guar gum	80–90% in weeks	Eco-friendly, renewable source
Gellan gum	85–95% in 2–4 weeks	Supports waste reduction initiatives

12.7 Water Conservation and Reduced Water Footprint

Water conservation is another key sustainability goal, as freshwater resources become increasingly scarce in many regions worldwide. The beauty and personal care industry, in particular, often uses large amounts of water in the formulation of products. Mucilages and gums can help reduce water consumption in both the **production process** and **end use**.

12.7.1 Goal: Conserving Water in Production and End Use

Many mucilage-based ingredients have inherent **high water retention** properties. This enables them to reduce the amount of water needed for their application, as they help retain moisture in both skin and hair care products (Table 12.2).

12.7.2 Data

- **Xanthan Gum:** Known for its excellent water retention, xanthan gum can absorb up to 10 times its weight in water, making it an ideal ingredient in moisturizers and hair care formulations. This high absorption capacity reduces the need for frequent application of products and minimizes water usage.
- **Water Footprint:** Products containing mucilages and gums have been shown to have a 30–40% lower water footprint than those made with synthetic ingredients such as polyethylene or silicone-based formulations.
- **Aloe Vera Gel:** Aloe vera is another mucilage-rich plant that has an extremely high moisture content, with its gel containing up to 98% moisture. The incorporation of aloe vera into dry skin care formulations helps to conserve water while improving hydration.

12.7.3 Impact

By using mucilages and gums with high water retention capacity, personal care and cosmetics companies can significantly reduce water usage in their products. This contributes to overall water conservation and reduced water footprint, aligning with sustainability goals focused on preserving vital resources.

Table 12.2: Water footprint reduction of mucilage.

Ingredient	Water retention capacity	Water footprint reduction
Xanthan gum	High (absorbs up to 10× weight)	30–40% lower than synthetic alternatives
Guar gum	High	Significant water conservation potential
Aloe vera gel	Extremely high (up to 98%)	Significant in formulations for dry skin care

12.8 Carbon Footprint Reduction

The **carbon footprint** of products refers to the amount of greenhouse gas emissions produced during the production, processing, and transportation of materials. As industries strive to reduce their carbon emissions, mucilages and gums offer a promising alternative to traditional synthetic ingredients, which are often more energy-intensive to produce.

12.8.1 Goal: Reducing the Carbon Emissions Associated with Manufacturing

The carbon footprint associated with synthetic thickeners and emulsifiers is often significant due to the energy-intensive processes involved in their production. Mucilages and gums, however, have a lower **energy requirement** for both **cultivation** and **processing** (Table 12.3).

12.8.2 Data

- **Guar Gum:** The cultivation of guar gum has been shown to have a 60% lower carbon footprint compared to synthetic thickeners. This reduction is largely attributed to the low-energy requirements of guar gum cultivation and its minimal need for synthetic fertilizers and pesticides.
- **Okra Gum:** The production of okra gum and other mucilage-based ingredients can be 3–5 times less energy-intensive than the production of synthetic emulsifiers such as polyethylene glycol (PEG) derivatives.
- **Aloe Vera Gel:** Aloe vera cultivation has a low carbon footprint, partly due to the plant's ability to grow in arid regions with minimal water, making it a sustainable crop for carbon reduction in farming systems.

12.8.3 Impact

By incorporating mucilages and gums into formulations, companies can significantly reduce the **carbon emissions** associated with product manufacturing. This contributes to broader sustainability goals related to climate change mitigation and reducing the carbon impact of industrial activities.

Table 12.3: Impact on greenhouse gas emissions of mucilage.

Ingredient	Carbon footprint reduction	Impact on greenhouse gas emissions
Guar gum	60% lower emissions	Supports carbon sequestration in soil
Okra gum	3–5× less energy-intensive	Helps reduce greenhouse gas emissions
Aloe vera	Low carbon footprint	Sustainable farming practices

12.9 Waste Reduction Through Circular Economy Practices

The concept of the **circular economy** revolves around reusing, recycling, and repurposing materials to reduce waste. Mucilages and gums play a crucial role in this model by enabling the **utilization of by-products** from their extraction process.

12.9.1 Goal: Minimizing Waste by Repurposing By-Products

After mucilage is extracted from plants, significant **residues** remain. In the past, these by-products were often discarded as waste, but now, companies are finding innovative ways to repurpose these materials, contributing to the circular economy.

12.9.2 Data

- **Okra Gum:** The waste products from okra gum extraction are being repurposed into biodegradable packaging materials, such as biofilms. These films are fully biodegradable and compostable, thus significantly reducing landfill waste.
- **Acacia Gum:** By-products from acacia gum extraction are being used to create bio-based materials, such as bio-plastics. These alternatives to petroleum-based plastics are not only eco-friendly but also help reduce landfill waste by 25–30%.
- **Guar Gum:** The by-products of guar gum production are being used for organic fertilizers and animal feed, contributing to waste repurposing and reducing the reliance on synthetic fertilizers and animal feed sources.

12.9.3 Impact

By repurposing the waste products from mucilage and gum extraction, industries contribute to the development of a more sustainable **circular economy**, where materials are reused, and waste is minimized. This not only reduces environmental impact but also creates new revenue streams for businesses (Table 12.4).

Table 12.4: Impact on waste reduction potential of mucilage.

Ingredient	By-product utilization	Waste reduction potential
Okra gum	Used in biodegradable packaging, biofilms	Reduces landfill waste by 25–30%
Acacia gum	Used in bio-based materials	Supports circular economy principles
Guar gum	Used for organic fertilizers, animal feed	Waste repurposing potential

12.10 Promoting Biodiversity and Regenerative Agriculture

Regenerative agriculture is a method of farming that focuses on restoring soil health, enhancing biodiversity, and reducing environmental damage. Many mucilage-producing plants are integral to **regenerative farming systems** that promote these goals.

12.10.1 Goal: Supporting Biodiversity and Promoting Regenerative Farming Practices

Plants like **acacia, guar, aloe vera**, and **okra** are often grown in **regenerative agriculture** systems that enhance soil health, support biodiversity, and contribute to climate change mitigation (Table 12.5).

12.10.2 Data

- **Acacia Gum:** Acacia trees, which produce gum, are grown in semi-arid regions and are key to preventing soil erosion while improving soil fertility. These trees also enhance biodiversity by supporting a variety of plant and animal species.
- **Aloe Vera Gel:** Aloe vera cultivation improves soil carbon sequestration by up to 50%, which helps mitigate climate change. Aloe plants also contribute to sustainable land management by thriving in dry and semi-arid conditions with minimal water input.
- **Okra Gum:** Okra is often grown in agroforestry systems, which foster diverse ecosystems while providing a renewable source of gum. This practice not only

contributes to biodiversity but also supports minimal water usage and soil regeneration.

12.10.3 Impact

The promotion of regenerative agriculture through mucilage and gum production helps enhance soil health, prevent erosion, and support biodiversity, which are key elements of a sustainable agricultural system.

Table 12.5: Impact on biodiversity of mucilage.

Ingredient	Role in regenerative agriculture	Impact on biodiversity
Acacia gum	Grown in semiarid regions, supports soil health	Prevents soil erosion, improves fertility
Aloe vera	Enhances soil carbon sequestration	Supports sustainable land management
Okra gum	Grown with minimal water usage, supports agroforestry	Encourages diverse ecosystems

12.11 Renewable and Ethical Sourcing

Mucilages and gums are often sourced from renewable plant sources that have a positive social and economic impact on the communities that cultivate them.

12.11.1 Goal: Ensuring Ethical Sourcing and Supporting Local Communities

Many mucilage-producing plants are harvested from regions that benefit local economies and communities. This makes mucilage production not only environmentally sustainable but also **socially beneficial**.

12.11.2 Data

- **Guar Gum:** The global market for guar gum is primarily based in India, where it provides livelihood to over 2 million farmers. The crop is a key agricultural product, especially in rural areas, contributing to local economies and providing social stability.

- **Acacia Gum:** Sourced from wild-harvested trees in Africa, acacia gum harvesting supports local economies by providing a source of income to communities in semi-arid regions. It also ensures ecosystem balance, as acacia trees are integral to the local environment.
- **Aloe Vera Gel:** Aloe vera is often grown in agroforestry systems, where it supports both local economies and sustainable farming practices. These systems promote biodiversity and ensure that communities benefit from environmentally sustainable crops (Table 12.6).

12.11.3 Impact

Ethical sourcing of mucilage-producing plants ensures that communities in developing regions benefit economically while promoting sustainable land use practices.

Table 12.6: Impact on community benefits of mucilage.

Ingredient	Ethical sourcing Impact	Community benefit
Guar gum	Supports 2 million farmers	Provides local livelihoods
Acacia gum	Wild-harvested, ethical sourcing	Supports local economies in Africa
Aloe vera	Grown in agroforestry systems	Promotes sustainable farming

12.12 Reduction of Plastic and Synthetic Chemicals in Products

The beauty industry is increasingly moving away from petroleum-based products and plastics. Mucilages and gums play a central role in this shift by offering natural alternatives to synthetic chemicals and plastics in cosmetics and personal care formulations.

12.12.1 Goal: Minimizing the Use of Synthetic Chemicals and Plastics

Mucilage-based emulsifiers and stabilizers serve as **natural alternatives** to synthetic thickeners, emulsifiers, and stabilizers, reducing the dependence on petrochemical-derived substances (Table 12.7).

12.12.2 Data

- **Xanthan Gum:** Used as a natural emulsifier and stabilizer, xanthan gum helps replace synthetic stabilizers in cosmetics, reducing the use of petrochemical-derived ingredients.
- **Guar Gum:** A natural thickener, guar gum offers an eco-friendly alternative to petroleum-based gels used in cosmetics and personal care products.
- **Aloe Vera Gel:** Aloe vera gel serves as a hydrating alternative to synthetic gels in skin care products, replacing petroleum-based gels with a natural, sustainable option.

12.12.3 Impact

By replacing synthetic chemicals with mucilage and gum-based alternatives, companies can reduce their environmental impact and dependence on plastics and petrochemicals, supporting the transition to more sustainable products.

Table 12.7: Impact of mucilage on reducing plastics.

Ingredient	Role in reducing plastics	Alternative uses
Xanthan gum	Natural emulsifier and stabilizer	Replaces synthetic stabilizers
Guar gum	Natural thickener, eco-friendly	Replaces petroleum-based gels
Aloe vera gel	Hydrating gel alternative	Replaces synthetic gels in skin care

Mucilages and gums represent an important category of natural ingredients that can significantly contribute to achieving sustainability goals across multiple industries. From biodegradability to water conservation, carbon footprint reduction, and waste minimization, these plant-based substances provide eco-friendly alternatives to synthetic ingredients.

12.13 Future Directions for Research, Industry, and Global Impact

12.13.1 Advancing Biotechnological Innovation for Improved Production

One of the most promising areas for future research is the **biotechnological advancement** of mucilage and gum production. Traditional methods of extracting gums from plants involve significant land use, water consumption, and energy input. However, the adoption of **biotechnology** could streamline the production process, reduce environmental impacts, and even increase yields.

Research into **fermentation-based production systems** is already underway, with studies showing that **microbial fermentation** can be a more efficient and scalable method for producing gums like **xanthan, guar**, and **gellan**. By genetically engineering bacteria and other microorganisms to produce these substances, the need for large-scale agricultural land could be minimized. This has the potential to drastically reduce the **carbon footprint** of gum production while improving the **sustainability** of the supply chain.

Furthermore, microbial fermentation could help reduce the use of harmful agricultural chemicals, as bacteria and fungi can often be cultivated with fewer inputs than conventional crops. With further development in **synthetic biology** and **genetic engineering**, it is plausible that the **biotechnological production** of mucilages could become as commonplace as that of industrial chemicals, offering a more **sustainable and scalable** alternative to traditional agricultural systems.

12.13.2 Exploring New Plant Sources and Diversification

While many industries have primarily relied on a few staple sources of mucilage and gums, such as **guar, xanthan**, and **acacia**, future research will likely involve exploring **under-researched plant species**. Indigenous and tropical plants may hold the key to new gums with unique properties, such as higher **water retention**, better **emulsification** characteristics, or improved **biodegradability**.

For instance, researchers could explore **tropical species** like **okra, cassava**, and **flax**, which may offer sustainable alternatives to conventional gum-producing plants. Additionally, the study of plants in diverse climatic regions, especially those that thrive in **semiarid** or **arid** environments, may uncover new sources of mucilages that are well-suited to **climate-resilient** farming systems.

One specific area of interest is the **search for gums from drought-resistant plants**. These species could provide an alternative to crops that require significant irrigation and inputs, making their inclusion in agricultural systems highly sustainable. Moreover, the exploration of plants from **indigenous ecosystems** may uncover unique properties that can create new applications in fields ranging from **biodegradable packaging** to **advanced wound healing materials**.

12.13.3 Improved Extraction Technologies

The current methods for extracting mucilage and gums are often energy-intensive and require large amounts of water. Future research into **greener extraction technologies** is essential for improving sustainability. **Cold-extraction** methods, for example, which do not require the use of high temperatures, could drastically reduce energy consumption. Similarly, the development of **supercritical fluid extraction** or

enzymatic extraction techniques may offer more efficient and environmentally friendly alternatives to traditional extraction methods.

Furthermore, **nanotechnology** has the potential to improve the **purity** and **quality** of mucilages, allowing for more efficient separation and concentration of gums. This would lead to higher-value products with more specialized applications, such as **biodegradable films, emulsifiers**, and **pharmaceutical delivery systems**.

Research into improving **yield efficiency** is also a significant area of interest. Maximizing the amount of gum that can be harvested from a given plant will not only increase the economic feasibility of mucilage production but also help meet the growing demand for these sustainable ingredients.

12.13.4 Understanding Mucilage's Functional Properties

Mucilages and gums are primarily used for their **viscosity-modifying** and **emulsifying properties**, but their potential goes far beyond these traditional applications. Future research will likely focus on better understanding the **functional properties** of mucilage and gums, particularly their **biochemical compositions** and interactions with other ingredients. This can unlock new applications in industries such as **biomedical engineering, food packaging**, and **environmental remediation**.

For example, mucilages from plants like **okra** and **aloe vera** have been used in **wound healing** due to their **moisture-retentive** and **anti-inflammatory** properties. However, these benefits could be further enhanced through **scientific studies** that explore their ability to deliver **active compounds** like vitamins and minerals directly to the skin or other tissues.

Moreover, the understanding of **gums as prebiotics** and their role in **gut health** has just begun to gain attention. Plant gums such as **guar gum** and **acacia gum** have been shown to have **prebiotic effects**, promoting the growth of beneficial gut bacteria. More research into the **gut microbiome** and the interactions between mucilages and gut flora could lead to new **functional food** products.

12.13.5 Circular Economy and Waste Utilization

Future research will increasingly focus on the **circular economy** of mucilage and gum production. The by-products of gum extraction, such as plant residues and waste materials, often end up as **agricultural waste** or **industrial by-products**. However, innovative research could lead to the **repurposing** of these materials into new value-added products.

For instance, research into **biodegradable packaging** made from gum by-products like **biofilms** or **bio-plastics** is an exciting avenue for reducing environmental waste. Additionally, by-products could be utilized as **fertilizers, biofuels**, or **ani-**

mal feed, closing the loop in the production process and contributing to sustainable waste management systems.

Through research into **waste-to-value technologies**, mucilage and gum-producing industries could minimize their ecological footprint while adding value to materials that would otherwise be discarded.

12.14 Evolving Industry Trends

12.14.1 Eco-friendly and Sustainable Product Formulations

As consumers become more environmentally conscious, there is an increasing demand for products that are natural, sustainable, and biodegradable. The beauty and personal care industries are already responding to this trend by incorporating more plant-based ingredients like gums and mucilages into their formulations. Going forward, it is expected that the trend toward green beauty will only continue to grow, with more brands prioritizing ingredients that are ethically sourced and sustainably produced.

The food industry is also making strides in adopting plant-based gums, particularly for their emulsifying, thickening, and stabilizing properties. As the global shift toward plant-based diets continues, the demand for natural ingredients that provide texture, consistency, and preservation without synthetic additives will increase.

To capitalize on these trends, industries will need to establish transparent supply chains, where the origin of mucilage and gum-based ingredients is traceable, and sustainability certifications are clearly communicated to consumers.

12.14.2 Regulatory and Certification Standards

As the market for plant-based gums continues to expand, **regulatory bodies** will likely introduce more stringent standards for the sourcing and production of these ingredients. This will ensure that products are not only effective but also **ethically produced** and **environmentally responsible**.

For instance, **organic certification** and **fair trade** certification could become increasingly important for manufacturers. Companies will need to comply with these regulations to gain consumer trust and remain competitive in the market. Additionally, there will likely be increased scrutiny of **supply chains** to ensure that production practices are aligned with sustainability goals, including the use of **regenerative agricultural techniques** and **renewable energy**.

12.15 Global Impact of Mucilage and Gums

12.15.1 Environmental Impact on Global Sustainability Goals

The widespread adoption of mucilages and gums in product formulations can contribute significantly to the **global sustainability agenda**. From reducing reliance on **petroleum-based** and **synthetic chemicals** to promoting the use of **biodegradable alternatives**, mucilage and gum-based products support many of the **UN Sustainable Development Goals (SDGs)**, including:

- **SDG 12 (Responsible Consumption and Production):** The use of natural, biodegradable gums reduces the environmental burden of synthetic chemicals and supports a more sustainable **circular economy**.
- **SDG 13 (Climate Action):** Many mucilage-producing plants, such as **acacia** and **guar**, help sequester **carbon** and mitigate **climate change** by promoting healthy soil systems.
- **SDG 15 (Life on Land):** The use of mucilage-producing plants in **agroforestry** systems can help restore **soil health**, reduce erosion, and improve biodiversity, contributing to the preservation of terrestrial ecosystems.

12.15.2 Socioeconomic Impact on Local Communities

Mucilage and gum production has a profound socioeconomic impact, particularly in developing countries where many of these plants are cultivated. For example, guar gum production supports over 2 million farmers in India, providing vital income and improving the livelihoods of rural communities.

As global demand for natural and sustainable products grows, the economic importance of mucilage and gum-producing plants will increase. By promoting ethical sourcing and fair-trade practices, these industries can further empower local communities, especially women and smallholder farmers, by creating new economic opportunities and supporting local economies.

Appendices

Appendix A: Glossary of Key Terms

- **Natural polymers**: Polymers derived from renewable resources, such as plants, animals, and microorganisms, which can naturally degrade over time.
- **Mucilage**: A viscous substance found in plants, algae, and microorganisms, often used in food, pharmaceuticals, and other industrial applications.
- **Gums**: A group of natural, hydrophilic polysaccharides that are used as thickeners, stabilizers, and emulsifiers in various industries.
- **Biodegradability**: The ability of a material to be broken down by microorganisms and returned to the natural environment without causing harm.
- **Rheology**: The study of the flow and deformation of matter, particularly how materials behave under stress.
- **Spectroscopy**: A technique used to measure the interaction of light with matter, which can provide information about the chemical structure of a material.
- **Microscopy**: The use of microscopes to examine the fine structure of materials, often at a cellular or molecular level.

Appendix B: Sources of Mucilage and Gums

1. **Plant-Based Sources**:
   - **Okra (*Abelmoschus esculentus*)**: Known for its mucilage, used in food and pharmaceuticals.
   - **Guar gum (*Cyamopsis tetragonoloba*)**: A leguminous plant gum, widely used in the food industry and as a thickener.
   - **Psyllium (*Plantago ovata*)**: The seeds contain mucilage used in laxatives and dietary supplements.
   - **Acacia gum (Gum Arabic)**: Used in food, pharmaceuticals, and cosmetics for its emulsifying properties.
2. **Algae-Based Sources**:
   - **Carrageenan (*Chondrus crispus*)**: Extracted from red seaweed, used in food and pharmaceuticals.
   - **Agar-agar**: A gelatinous substance derived from red algae, used as a gelling agent in microbiological media and food.
3. **Microbial Sources**:
   - **Xanthan gum (*Xanthomonas campestris*)**: A bacterial exopolysaccharide used in the food and cosmetic industries.
   - **Dextran**: Produced by certain bacteria, commonly used in the pharmaceutical and food industries.

https://doi.org/10.1515/9783111673509-013

Index

https://doi.org/10.1515/9783111673509-014